大数据与人工智能技术丛书

大数据专业英语教程

第2版

◎ 朱丹 朱宏峰 刘伟 编著

清華大学出版社
北京

内 容 简 介

本书是计算机、信息管理和大数据等相关专业的专业英语教材，选材广泛，覆盖大数据的数据挖掘、数据分析等各个方面，同时兼顾了相关发展热点。本书所选取的文章包括以下内容：大数据的基本概念、大数据的数据挖掘、大数据的数据分析、大数据的影响、大数据的商业价值、大数据在各个领域的应用以及大数据如何改变人们的生活等。每章所选用文章均来自国外网站，文章中出现的新词和专业术语也均有注释，每篇文章配有相应的习题和拓展阅读，以巩固学习效果。

本书可作为高等本科院校、高等专科院校大数据相关专业的英语教材，也可供从业人员自学和参考。

图书在版编目（CIP）数据

大数据专业英语教程 / 朱丹，朱宏峰，刘伟编著. —2 版. —北京：清华大学出版社，2023.12（2024.12重
（大数据与人工智能技术丛书）

ISBN 978-7-302-64493-4

Ⅰ. ①大… Ⅱ. ①朱… ②朱… ③刘… Ⅲ. ①数据处理 -英语 -教材 Ⅳ. ①TP274

中国国家版本馆 CIP 数据核字（2023）第 164911 号

责任编辑：赵 凯 李 晔
封面设计：刘 键
责任校对：郝美丽
责任印制：丛怀宇

出版发行：清华大学出版社
网 址：https://www.tup.com.cn，https://www.wqxuetang.com
地 址：北京清华大学学研大厦 A 座 邮 编：100084
社 总 机：010-83470000 邮 购：010-62786544
投稿与读者服务：010-62776969，c-service@tup.tsinghua.edu.cn
质 量 反 馈：010-62772015，zhiliang@tup.tsinghua.edu.cn
课 件 下 载：https://www.tup.com.cn，010-83470236
印 装 者：三河市铭诚印务有限公司
经 销：全国新华书店
开 本：185mm×260mm 印 张：9.75 字 数：240 千字
版 次：2018 年 9 月第 1 版 2023 年 12 月第 2 版 印 次：2024 年 12 月第 2 次印刷
印 数：1501～2300
定 价：49.00 元

产品编号：095951-01

第2版 前言
FOREWORD

随着时代的进步和社会的高速发展，互联网行业发展风起云涌，移动互联网、电子商务、物联网以及社交媒体的快速发展促使我们快速进入了大数据时代。大数据技术与应用相关专业前景相当广阔，大数据人才需求旺盛，统计显示，2015—2022年，全国共有715所高校成功备案“数据科学与大数据技术”本科专业。

大数据专业处于高速发展之中，国际化特征尤为明显，从业人员必须提高专业英语水平，以便及时获得最新、最先进的专业知识。从某种意义上说，专业英语的水平决定了专业技能的水平。了解和掌握一些大数据专业英语是非常有必要的，因此，几乎所有开设大数据专业的高校都开设了相应的专业英语课程。

本书的内容包括大数据的基本概念、大数据的数据挖掘、大数据的数据分析、大数据的影响、大数据的商业价值、大数据在各个领域的应用以及大数据如何改变人们的生活等。文章均选自国外知名网站，具有一定的知识性和实用性；New Words and Expressions 给出文章中出现的新词，读者由此可以扩充词汇量；Terms 对文中出现的专业术语进行解释；Comprehension 针对文章进行练习，巩固学习效果；Answers 给出参考答案，读者可对照检查学习效果；参考译文帮助读者理解文章大意；“常用大数据词汇中英文对照表”供读者记忆单词和查询之用。

本书可作为大数据专业相关课程教材，英语专业及计算机专业的选修教材，各类院校大数据和相关专业的参考书，也可作为各类计算机从业人员或有志投身于大数据领域的人士的自学书籍。

本书第1章（Chapter 1）至第8章（Chapter 8）由朱丹编写，第9章（Chapter 9）及“常用大数据词汇中英文对照表”由朱宏峰编写，第10章（Chapter 10）和第11章（Chapter 11）由刘伟编写。全书由朱丹统稿。

本书文章节选自互联网，在此向文章原作者表示感谢，由于编者水平有限，书中难免存在不足之处，敬请读者不吝指正。

编者

2023年3月

目 录
CONTENTS

Chapter 1

What is Big Data

Text A

Big data is being generated by everything around us at all times. Every digital process and social media exchange produces it. Systems, sensors and mobile devices transmit it. Big data is arriving from multiple sources at an alarming velocity, volume and variety. To extract meaningful value from big data, you need optimal processing power, analytics capabilities and skills.

Big data is a relative term describing a situation where the volume, velocity and variety of data exceed an organization's storage or compute capacity for accurate and timely decision making.

Some of this data is held in transactional data stores — the byproduct of fast-growing online activity. Machine-to-machine interactions, such as metering, call detail records, environmental sensing and RFID systems, generate their own tidal waves of data. All these forms of data are expanding, and that is coupled with fast-growing streams of unstructured and semi structured data from social media.

However, big data is defined less by volume — which is a constantly moving target — than by its ever-increasing variety, velocity, variability and complexity.

Variety. Up to 85 percent of an organization's data is unstructured — not numeric — but it still must be folded into quantitative analysis and decision making. Text, video, audio and

New Words and Expressions

sensor/ˈsensə(r)/ *n.*
传感器

transmit/trænsˈmɪt/ *v.*
播送，发射，传送（信号）

velocity/vəˈlɒsəti/ *n.*
速度；速率

extract/ɪkˈstrækt/ *v.*
提取

optimal/ˈɒptɪməl/ *adj.*
最优的，最佳的；优化的

analytics/ˌænəˈlɪtɪks/ *n.*
分析，逻辑分析的方法

exceed/ɪkˈsiːd/ *v.*
超过，胜过

transactional/trænˈzækʃənəl/ *adj.*
交易的，业务的

metering/ˈmiːtərɪŋ/ *n.*
测量（法），测定

tidal /ˈtaɪdl/ *adj.*
潮汐的，潮水的

numeric /njuːˈmerɪk/ *adj.*
数字的，数值的

quantitative/ˈkwɒntɪtətɪv/ *adj.*
定量的，数量（上）的

other unstructured data require different architecture and technologies for analysis.

Velocity. Thornton May says, "Initiatives such as the use of RFID tags and smart metering are driving an ever greater need to deal with the torrent of data in near-real time. This, coupled with the need and drive to be more agile and deliver insight quicker, is putting tremendous pressure on organizations to build the necessary infrastructure and skill base to react quickly enough."

Variability. In addition to the speed at which data comes your way, the data flows can be highly variable — with daily, seasonal and event-triggered peak loads that can be challenging to manage.

Complexity. Difficulties dealing with data increase with the expanding universe of data sources and are compounded by the need to link, match and transform data across business entities and systems. Organizations need to understand relationships, such as complex hierarchies and data linkages, among all data.

A data environment can become extreme along any of the above dimensions or with a combination of two or all of them at once. However, it is important to understand that not all of your data will be relevant or useful. Organizations must be able to separate the wheat from the chaff and focus on the information that counts — not on the information overload.

What is changing in the realm of big data?

Big data is changing the way people within organizations work together. It is creating a culture in which business and IT leaders must join forces to realize value from all data. Insights from big data can enable all employees to make better decisions — deepening customer engagement, optimizing operations, preventing threats and fraud, and capitalizing on new sources of revenue. But escalating demand for insights requires a fundamentally new approach to architecture, tools and practices.

Competitive advantage: Data is emerging as the world's newest resource for competitive advantage.

Decision making: Decision making is moving from the elite few to the empowered many.

Value of data: As the value of data continues to grow, current systems won't keep pace.

New Words and Expressions

torrent/ˈtɒrənt/ *n.*
奔流

agile/ˈædʒaɪl/ *adj.*
灵活的，机敏的

peak loads
峰值负荷

entity/ˈentəti/ *n.*
实体

hierarchy/ˈhaɪərɑːki/ *n.*
[计]分层，层次，等级制度

linkage/ˈlɪŋkɪdʒ/ *n.*
联系，连接

separate the wheat from th chaff
分清良莠

realm/relm/ *n.*
领域，范围

optimize/ˈɒptɪmaɪz/ *v.*
优化，完善

fraud/frɔːd/ *n.*
诈骗（罪）

revenue/ˈrevənjuː/ *n.*
（公司的）收益，（政府的）税收

escalate/ˈeskəleɪt/ *v.*
（使）增强，（使）扩大

elite/iˈliːt/ *n.*
掌权人物，精英

empower/ɪmˈpaʊər/ *v.*
给（某人）做…的权力，授权

How can you realize the greatest value from big data?

New skills are needed to fully harness the power of big data. Though courses are being offered to prepare a new generation of big data experts, it will take some time to get them into the workforce. Meanwhile, leading organizations are developing new roles, focusing on key challenges and creating new business models to gain the most from big data.

- *Discover the new role of data scientist*

Gartner finds that by 2015, the demand for data and analytics resources will reach 4.4 million jobs globally, but only one-third of those jobs will be filled. The emerging role of data scientist is meant to fill that skills gap.

- *Be proactive about privacy, security and governance*

While big data can provide significant value, it also presents significant risk. Organizations must be proactive about privacy, security and governance to ensure all data and insights are protected and secure.

- *Create new business models with big data*

From data-driven marketing and ad targeting to the connected car, big data is fueling product innovation and new revenue opportunities for many organizations.

Employ the most effective big data technology

To gain the competitive advantage that big data holds, you need to infuse analytics everywhere, make speed a differentiator, and exploit value in all types of data. This requires an infrastructure that can manage and process exploding volumes of structured and unstructured data — in motion as well as at rest — and protect data privacy and security.

Big data technology

Big data technology must support search, development, governance and analytics services for all data types — from transaction and application data to machine and sensor data to social, image and geospatial data, and more.

- *Systems*

Your infrastructure must capitalize on real-time information flowing through your organization. It must be optimized for analytics to respond dynamically — with automated business processes, better agility and improved economics — to the increasing demands of big data.

New Words and Expressions

harness/ˈhɑːnɪs/ *vt.*
利用

proactive/ˌprəʊˈæktɪv/ *adj.*
积极主动的，前摄的

governance/ˈgʌvənəns/ *n.*
管理，统治

innovation/ˌɪnəˈveɪʃn/ *n.*
改革，创新

infuse/ɪnˈfjuːz/ *vt.*
注入，灌输

differentiator/dɪfəˈrenʃɪeɪtə/ *n.*
区分者，微分器

exploit/ɪkˈsplɔɪt/ *vt.*
开拓，开采

agility /əˈdʒɪlətɪ/ *n.*
敏捷，灵活

- *Privacy*

To protect your reputation and brand, your platform must comprise stringent policies and practices around privacy and data protection, safeguarding all of the data and insights on which your business relies.

- *Governance*

The right platform instills trust, so you can act with confidence. It controls how information is created, shared, cleansed, consolidated, protected, maintained, retired and integrated within your enterprise.

- *Storage*

To achieve economies and efficiencies, you must run certain analytics close to the data, while it is in motion. But for data you elect to store, your infrastructure must embody a defensible disposal strategy that reduces the run rate of storage, legal expense and risk.

- *Security*

As you infuse analytics into your organization, data security becomes more central to your competitive advantage profile. Your infrastructure must have strong security measures built in to guard your organization against internal and external threats.

- *Cloud*

To relieve the pressure that big data is placing on your IT infrastructure, you can host big data and analytics solutions on the cloud. Achieve the scalability, flexibility, expandability and economics that will provide competitive advantage into the future.

New Words and Expressions

reputation/ˌrepju'teɪʃn/ *n.* 名声

comprise/kəm'praɪz/ *vt.* 包含，包括

safeguard/'seɪfgɑ:d/ *vt.* 防护，保卫

instill /ɪn'stɪl/ *vt.* 逐渐使某人获得（某种可取的品质）

cleanse/klenz/ *vt.* 净化，清洗

consolidate/kən'sɒlɪdeɪt/ *vt.* 统一，合并

integrate/'ɪntɪgreɪt/ *vt.* 使一体化

efficiency/ɪ'fɪʃnsi/ *n.* 功效，效率

defensible/dɪ'fensəbl/ *adj.* 能防御的

scalability/skeɪlə'bɪlɪtɪ/ *n.* 可量测性

flexibility/ˌfleksə'bɪlətɪ/ *n.* 灵活性

expandability/ɪks'pændəbɪlɪtɪ/ *n.* 扩展性

Note:

The text is adapted from the website:

http://www.ibm.com/big-data/us/en/.

Comprehension

Blank Filling

1. Big data is being generated by everything around us at all times. Every__________ and __________ produces it.
2. Big data is arriving from multiple sources at an alarming ________, ________ and ________. To extract meaningful value from big data, you need ________ power, ________capabilities and skills.

3. Big data is a relative term describing a situation where the volume, velocity and variety of data exceed an organization's _______ or _______ capacity for accurate and timely _______________.
4. Insights from big data can enable all employees to make better decisions — deepening __________, optimizing _______, preventing__________, and capitalizing on new sources of________.
5. Meanwhile, leading organizations are developing _______, focusing on key _______ and creating new _________ to gain the most from big data.
6. To gain the competitive advantage that big data holds, you need to infuse _______ everywhere, make speed a differentiator, and exploit ________ in all types of data.
7. Big data technology must support _______, _________, _________ and _________ services for all data types — from ___________ data to __________machine and sensor data to ________________ data, and more.
8. To relieve the pressure that big data is placing on your IT infrastructure, you can host big data and analytics solutions on the _______.
9. IBM data scientists break big data into four dimensions:__________________________.

Content Questions

1. What is the definition of big data?
2. What are the characteristics of big data?
3. What is the background of big data?
4. What does the big data technology do?
5. What is the value of digging big data?

Answers

Blank Filling

1. digital process; social media exchange
2. velocity; volume; variety; optimal processing; analytics
3. storage; compute; decision making
4. customer engagement; operations; threats and fraud; revenue
5. new roles; challenges; business models
6. analytics; value
7. search; development; governance; analytics; transaction and application; social; image and geospatial
8. cloud
9. volume, variety, velocity and veracity

Content Questions

1. "Big data" is a massive, high-growth and diversified information asset that requires a new processing model which have greater decision-making power, insight into discovery

and process optimization capabilities.

2. Volume, Variety, Velocity, Value.
3. Big data is being generated by everything around us at all times. Every digital process and social media exchange produces it. Systems, sensors and mobile devices transmit it. Big data is arriving from multiple sources at an alarming velocity, volume and variety.
4. Big data technology must support search, development, governance and analytics services for all data types — from transaction and application data to machine and sensor data to social, image and geospatial data, and more.
5. The value of digging big data is similar to sandy gold rush, digging sparse but valuable information from massive amounts of data.

参考译文

我们身边的一切每时每刻都在产生大量的数据。每个数字流程和社交媒体的互动都会产生数据。这些数据通过系统、传感器和移动设备进行传输。大数据源于众多数据源，其产生速度、数据量和多样性都迅速增长。要从大数据中提取有意义的信息价值，需要最优的处理能力、分析能力和技术。

大数据是描述数据量、数据获得速度和数据多样性的名词术语，大数据受限于硬件设施，从而使一些公司在存储空间、计算资源方面不能提供准确、实时的分析结果。

其中一些数据存储在基于事务模型的数据库中——这是快速增长的在线活动的副产品。机器对机器的交互，如计量、通话细节记录、环境感测和 RFID 系统，产生自己的大量数据。所有这些形式的数据正在快速增长，同时，来自社交媒体的非结构化和半结构化数据也在飞速增长。

然而，大数据并不定义为“量”，而指的是不断增加的数据种类、数据产生速度、数据多样性和复杂性。

多样性。组织中的数据有高达 85%的部分是非结构化的（非数字形式），但其必须转化为数字形式，以用于定量分析和决策。文本、视频、音频和其他非结构化数据需要不同的架构和技术进行分析。

高速。Thornton May 指出：“射频识别（RFID）和智能计量等新技术的使用，正推动人们越来越迫切地需要实时地处理海量数据，并更加敏捷快速地给出见解，这给公司带来巨大的压力，必须建立必要的基础设施和技能库，以迅速作出反应。”

变化性。除了数据传输的速度之外，数据流可能是高度可变的，日常的、季节性的和事件触发的峰值负载都可能对管理带来挑战。

复杂性。随着数据源不断增多，处理数据的难度越来越大，需要在业务实体和系统之间链接、匹配和转换数据。组织需要了解所有数据之间的关系，例如复杂的层次结构和数据链接。

数据环境可以在上述任何方面变得极端，更不用说上述几方面还可能组合出现。但是，重要的是要了解并不是所有的数据都是相关的或有用的。为了应对信息过载，组织需要具备筛选重点信息并关注有效信息的能力。

大数据领域的变化是什么？

大数据正在改变组织内部人员的合作方式。它正在创造一种文化，企业和IT领导者必须共同努力，使大数据的价值得以体现。来自大数据的分析结论可以使所有员工做出更好的决策——深化客户参与，优化运营，防止威胁和欺诈，以及探索新的收入来源。但是，由于洞察力需求的不断增长，需要一种全新的方法来构建、使用和实践。

竞争优势：数据正在成为世界上最新的竞争优势资源。

决策过程：决策正在从精英阶层转向被赋予权力的许多人。

数据价值：随着数据价值的不断增长，目前的系统将不能保持同步。

如何从大数据中获得最大价值？

大数据的能量需要新的技术来发掘，虽然一些课程正在培养新一代大数据专家，但需要一段时间他们才能工作。同时，领先企业正在发挥新的作用，重点关注重大挑战，创造新的商业模式，从大数据中获取最大收益。

- 发现数据科学家的新角色

Gartner 公司发现，到 2015 年，对数据和分析资源的需求将在全球创造 440 万个工作岗位，但只有三分之一的岗位得到落实。数据科学家这一新兴角色意在填补这一技能缺口。

- 积极主动关注隐私、安全和管理

虽然大数据可以提供重要的价值，但也存在重大风险。公司机构必须积极主动地了解隐私、安全和管理，以确保所有数据和分析得到妥善保护。

- 使用大数据创建新业务模式

从数据驱动的营销、广告定向投放到联网汽车，大数据推动了许多公司的产品创新和新的收入机会。

采用最有效的大数据技术

为了获得大数据所具有的竞争优势，人们可以在任何客户端上输入分析数据，使速度成为一个产生区别的主要因素，并深度挖掘不同类型数据的价值。因此必须设计一个完善的基础架构，可以管理和处理以指数级增长的结构化和非结构化的数据量（包括静态数据与动态数据），同时保护数据的隐私和安全。

大数据技术

大数据技术必须支持所有数据类型的搜索、开发、管理和分析服务，从交易数据、应用程序数据到机器和传感器数据，以及社交化信息、图像和地理空间数据等。

- 系统

大数据基础设施必须利用流经公司组织的实时信息，同时它必须对数据分析进行优化以便动态响应，包括自动化业务流程、高便捷性和高性价比，以满足大数据日益增长的需求。

- 隐私

为了保护公司的声誉和品牌，大数据平台必须包含有关隐私和数据保护的严格策略和机制，保护公司业务所依靠的数据和未来规划。

- 管理

拥有可信度的完善大数据平台可以让用户或企业在使用时更加放心。它控制如何在企业中创建、共享、清理、整合、保护、维护、删除和集成信息。

- 存储

为了实现经济性和高效性，必须在运行过程中执行与数据关系密切的特定分析。但是，对于用户选择存储的数据，平台的基础架构必须体现出可防范的处置策略，从而减少运行存储系统的费用、法律费用和风险。

- 安全

当企业将分析数据上传到大数据平台时，数据安全将成为企业竞争优势的核心。大数据的基础架构必须具有强大的安全措施，以保护企业免受内部和外部威胁。

- 云

为了减轻大数据在IT基础设施上的压力，可以在云端托管大数据和分析解决方案，以实现可伸缩性、灵活性、可扩展性和经济性，为未来提供竞争优势。

Text B

Big data is increasingly becoming a factor in production, market competitiveness. Cutting-edge analysis technologies are making inroads into all areas of life and changing our day-to-day existence. Sensor technology, biometric identification and the general trend towards a convergence of information and communication technologies are driving the big data movement.

Huge challenges must be overcome if the benefits are to be leveraged effectively. Matters of concern alongside increasing volumes of data, varying data structures and real-time processing include data security, data privacy policies that are in urgent need of reform and the rising quality expectations of the stakeholders.

Using sensors, a multitude of data sets and specific algorithms, automatic predictions could soon be made about particular behavioral tendencies (and not just online) on the basis of simple correlations. The way in which people think about data and data analysis will gradually change as well, in addition to the technological possibilities.

Big data is more than just IT

Many decision-makers in all kinds of sectors have recognized that big data is no longer purely the preserve of IT. Big data is instead becoming a movement that brings together cutting-edge internet technologies and analysis techniques in order for large, extendable and above all differently structured data sets to be captured, stored and analyzed. This gives big data a broad, international dimension with different knowledge-based outcomes

New Words and Expressions

cutting-edge
前沿的

inroad /ˈɪnrəʊd/ *n.*
进展

convergence/kənˈvɜːdʒəns/ *n.*
会聚，聚集，收敛

dimension/daɪˈmenʃn/ *n.*
[数]次元，度，维

and expectations with regard to increasing growth and efficiency. But above all, big data provides scope for experimentation, innovation and creativity, offers a wealth of potential new data combinations and is therefore ideal for discovering unexpected correlations. It could be used to create new business models, products and services and to drive innovation.

The information management big data and analytics capabilities include:

Data Management & Warehouse: Gain industry-leading database performance across multiple workloads while lowering administration, storage, development and server costs; Realize extreme speed with capabilities optimized for analytics workloads such as deep analytics, and benefit from workload-optimized systems that can be up and running in hours.

Hadoop System: Bring the power of Apache Hadoop to the enterprise with application accelerators, analytics, visualization, development tools, performance and security features.

Stream Computing: Efficiently deliver real-time analytic processing on constantly changing data in motion and enable descriptive and predictive analytics to support real-time decisions. Capture and analyze all data, all the time, just in time. With stream computing, store less, analyze more and make better decisions faster.

Content Management: Enable comprehensive content lifecycle and document management with cost-effective control of existing and new types of content with scale, security and stability.

Information Integration & Governance: Build confidence in big data with the ability to integrate, understand, manage and govern data appropriately across its lifecycle.

New Words and Expressions

Stream Computing
流计算

Note:

The text is adapted from the website:

https://www-01.ibm.com/software/data/bigdata/.

参考译文

大数据越来越成为影响生产和市场竞争力的因素。先进的分析技术正在进入生活的各个方面，改变我们的日常生活。传感器技术、生物识别和信息通信技术融合的趋势正在驱动大数据快速发展。

要有效利用这些好处，必须克服巨大的挑战。大数据亟待解决的重要问题包括数据量的不断增长、数据结构的不断变化和需求的实时处理、数据的安全性、迫切需要改革的数据隐私策略以及利益相关者不断提高的期望值。

使用传感器、多种数据集和特定算法可以在简单的相关性基础上就特定的行为倾向（而不只是在线）做出自动预测。除了技术可能性之外，人们对数据和数据分析的思考方式也将逐渐改变。

各行业的利益相关者：大数据不仅仅是 IT

各行业的决策者都认识到，大数据不再仅仅是 IT 部门独占的领域。相反，大数据将汇集先进的互联网技术和分析技术，以便进行对大型的、可扩展的和各种不同结构的数据集的捕捉、存储和分析。这为大数据提供了广泛的国际应用空间、不同的知识成果和对效率日益增长的期望。但最重要的是，大数据为实验、创新和创造力的发展空间，提供了大量潜在的新数据组合，因此大数据分析是发现数据之间意外相关性的最佳方式。它可以用于创建新的商业模式、产品和服务，并能推动创新。

信息管理大数据和分析功能包括以下方面。

数据管理和仓库：在降低管理、存储、开发和服务器成本的同时，大数据可在多个工作负载下获得行业领先的数据库性能；通过对分析工作负载（如深度分析）进行优化的功能，实现极高的速度，并可从数小时内启动和运行的工作负载优化系统中获益。

Hadoop 系统：通过应用加速器、分析、可视化、开发工具、性能和安全功能，将 Apache Hadoop 的强大功能带入企业。

流计算：有效地为不断变化的运动数据提供实时分析处理，并支持描述性和预测性分析，以支持实时决策，无时无刻地捕捉并分析所有的数据。使用流计算，可以减少存储空间，做更多的分析，更快地做出更好的决策。

内容管理：通过规模、安全性和稳定性，对现有和新类型的内容进行成本效益的控制，实现全面的内容生命周期和文档管理。

信息集成和治理：建立对大数据的信心，并在整个周期中适当地集成、推断、管理和支配数据。

Chapter 2

Data Mining for Big Data

Text A

Data mining involves exploring and analyzing large amounts of data to find patterns for big data. The techniques came out of the fields of statistics and artificial intelligence (AI), with a bit of database management thrown into the mix.

Generally, the goal of the data mining is either classification or prediction. In classification, the idea is to sort data into groups. For example, a marketer might be interested in the characteristics of those who responded versus who didn't respond to a promotion.

These are two classes. In prediction, the idea is to predict the value of a continuous variable. For example, a marketer might be interested in predicting those who will respond to a promotion.

Typical algorithms used in data mining include the following:

Classification trees — A popular data-mining technique that is used to classify a dependent categorical variable based on measurements of one or more predictor variables. The result is a tree with nodes and links between the nodes that can be read to form if-then rules.

A tree showing survival of passengers on the Titanic ("sibsp" is the number of spouses or siblings aboard). The figures under the leaves show the probability of outcome and the percentage of observations in the leaf (shown in Figure 2-1).

Logistic regression[1] — A statistical technique that is a variant of standard regression but extends the concept to deal with

New Words and Expressions

statistic /stəˈtɪstɪk/ *n.*
统计，统计学

artificial intelligence
人工智能

promotion /prəˈməʊʃən/ *n.*
促销，推销；宣传

continuous variable
连续变量

dependent categorical variable
相关的分类变量

predictor variable
预测变量

classification. It produces a formula that predicts the probability of the occurrence as a function of the independent variables[2].

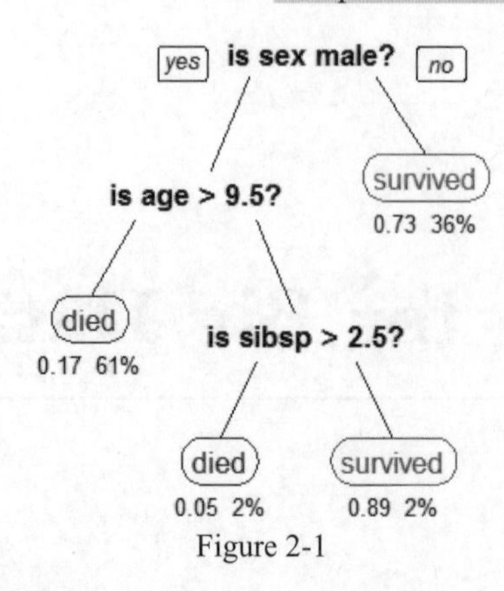

Figure 2-1

Neural networks[3] — A software algorithm that is modeled after the parallel architecture of animal brains. The network consists of input nodes, hidden layers, and output nodes. Each unit is assigned a weight. Data is given to the input node, and by a system of trial and error, the algorithm adjusts the weights until it meets a certain stopping criteria. Some people have likened this to a black-box approach[4].

Clustering[5] techniques like K-nearest neighbors[6] — A technique that identifies groups of similar records. The K-nearest neighbor technique calculates the distances between the record and points in the historical (training) data. It then assigns this record to the class of its nearest neighbor in a data set.

Here's a classification tree example. Consider the situation where a telephone company wants to determine which residential customers are likely to disconnect their service. The telephone company has information consisting of the following attributes: how long the person has had the service, how much he spends on the service, whether the service has been problematic, whether he has the best calling plan he needs, where he lives, how old he is, whether he has other services bundled together, competitive information concerning other carriers plans, and whether he still has the service.

Of course, you can find many more attributes than this. The last attribute is the outcome variable; this is what the software will use to classify the customers into one of the two groups — perhaps called stayers and flight risks.

New Words and Expressions

logistic regression
逻辑回归

variant/ˈveəriənt/ *n.*
变体，变形

independent variable
自变量

assign /əˈsaɪn/ *vt.*
分派，赋值

stopping criteria
停止准则

clustering/ˈklʌstərɪŋ/ *n.*
群集，聚类

K-nearest neighbors
K 最近邻（算法）

bundle/ˈbʌndəl/ *vt.&vi.*
收集，归拢，把…塞入

The data set is broken into training data and a test data set. The training data consists of observations (called attributes) and an outcome variable (binary in the case of a classification model) — in this case, the stayers or the flight risks.

The algorithm is run over the training data and comes up with a tree that can be read like a series of rules. For example, if the customers have been with the company for more than ten years and they are over 55 years old, they are likely to remain as loyal customers.

These rules are then run over the test data set to determine how good this model is on "new data." Accuracy measures are provided for the model. For example, a popular technique is the confusion matrix[7]. This matrix is a table that provides information about how many cases were correctly versus incorrectly classified.

If the model looks good, it can be deployed on other data, as it is available (that is, using it to predict new cases of flight risk). Based on the model, the company might decide, for example, to send out special offers to those customers whom it thinks are flight risks.

New Words and Expressions

flight risks
潜逃风险

binary/ˈbanəri/ *adj.*
二进制

confusion matrix
含混矩阵

deploy/dɪˈplɔɪ/ *vt.*
有效运用

Note:

The text is adapted from the website:

http://www.dummies.com/programming/big-data/ engineering/data-mining-for-big-data/.

Terms

1. Regression

In statistical modeling, regression analysis is a set of statistical processes for estimating the relationships between a dependent variable (outcome variable) and one or more independent variables (predictors).

回归：用一个或多个预测变量（predictor）来预测结果变量（outcome variable）值的统计分析。

2. Independent variables/dependent variables

The independent variable is the cause. Its value is independent of other variables in your study. The dependent variable is the effect. Its value depends on changes in the independent variable.

自变量/因变量：自变量是因，而因变量是果，自变量的发生在前，因变量的发生在后。自变量有时在不同的情境中亦被称为预测变量，而因变量则被称为效标变量。

3. Neural networks

Neural networks, also known as artificial neural networks (ANNs) or simulated neural networks (SNNs), are a subset of machine learning and are at the heart of deep learning algorithms. Their name and structure are inspired by the human brain, mimicking the way that biological neurons signal to one another. Neural networks rely on training data to learn and improve their accuracy over time.

神经网络：人工神经网络（Artificial Neural Network，ANN）简称神经网络，它是一种模仿动物神经网络行为特征，进行分布式并行信息处理的算法数学模型。这种网络依靠系统的复杂程度，通过调整内部大量节点之间相互连接的关系，从而达到处理信息的目的。

4. Black-box approach

Black-box testing assesses a system solely from the outside, without the operator or tester knowing what is happening within the system to generate responses to test actions. A black box refers to a system whose behavior has to be observed entirely by inputs and outputs.

黑盒方法：通常在分析一个开放系统时，采用典型的黑盒方法，系统的模拟数据作为黑盒输入，系统的反馈结果作为输出，而黑盒中的所有运作过程无法得知，如下图所示。黑盒方法主要针对系统的功能性测试。

而白盒方法则适用于已知系统的内部工作过程，可以通过测试证明每种内部操作是否符合设计规格要求，所有内部成分是否已经过检查。

5. Clustering

Clustering is the task of grouping a set of objects in such a way that objects in the same group (called a cluster) are more similar (in some sense) to each other than to those in other groups (clusters). It is a main task of exploratory data analysis, and a common technique for statistical data analysis, used in many fields, including pattern recognition, image analysis, information retrieval, bioinformatics, data compression, computer graphics and machine learning.

Cluster analysis can be achieved by various algorithms that differ significantly in their understanding of what constitutes a cluster and how to efficiently find them. Popular notions of clusters include groups with small distances between cluster members, dense areas of the data space, intervals or particular statistical distributions.

聚类：聚类的本质就是寻找联系紧密的事物，把它们区分出来。如果这些事物较少，人为地就可以简单完成这一目标。但是遇到大规模的数据时，人力就显得十分无力了。所以我们需要借助计算机来帮助寻找海量数据间的联系。

聚类过程中有一个关键的量，这个量就是标识两个事物之间的关联度的值，称为相关距离度量（distance metrics），相似性度量、皮尔逊相似性系数都是计算这种距离度量的方法。根据实际情况的不同，选择不同的适用的度量方法。这一点十分重要，直接影响聚类的结果是否符合实际需要和情况。

聚类是一个无监督学习（unsupervised learning）的过程，无须进行样本数据的训练。设计出适合的距离度量方法后，即可对目标数据集进行聚类。

6. K-Nearest Neighbors

In statistics, the K-Nearest Neighbors (KNN) algorithm is used for classification and regression. In both cases, the input consists of the k closest training examples in a data set. The output depends on whether KNN is used for classification or regression:

In KNN classification, the output is a class membership. An object is classified by a plurality vote of its neighbors, with the object being assigned to the class most common among its k nearest neighbors (k is a positive integer, typically small). If $k = 1$, then the object is simply assigned to the class of that single nearest neighbor.

In KNN regression, the output is the property value for the object. This value is the average of the values of k nearest neighbors.

K 最近邻（K-Nearest Neighbor，KNN）算法：是一个理论上比较成熟的方法，也是最简单的机器学习算法之一。该方法的思路是：如果一个样本在特征空间中的 k 个最相似（即特征空间中最邻近）的样本中的大多数属于某一个类别，则该样本也属于这个类别。KNN 算法中，所选择的邻居都是已经正确分类的对象。该方法在定类决策上只依据最邻近的一个或者几个样本的类别来决定待分样本所属的类别。KNN 算法虽然从原理上也依赖于极限定理，但在类别决策时，只与极少量的相邻样本有关。由于 KNN 算法主要靠周围有限的邻近的样本，而不是靠判别类域的方法来确定所属类别的，因此对于类域的交叉或重叠较多的待分样本集来说，KNN 算法较其他方法更为适合。

KNN 算法不仅可以用于分类，还可以用于回归。通过找出一个样本的 k 个最近邻居，将这些邻居的属性的平均值赋给该样本，就可以得到该样本的属性。更有用的方法是将不同距离的邻居对该样本产生的影响给予不同的权值（weight），如权值与距离成正比。

该算法在分类时有个主要的不足是，当样本不平衡时，如一个类的样本容量很大，而其他类样本容量很小时，有可能导致当输入一个新样本时，该样本的 k 个邻居中大容量类的样本占多数。因此可以采用权值的方法（和该样本距离小的邻居权值大）来改进。该算法的另一个不足之处是计算量较大，因为对每一个待分类的文本都要计算它到全体已知样本的距离，才能求得它的 k 个最近邻点。目前常用的解决方法是事先对已知样本点进行剪辑，事先去除对分类作用不大的样本。该算法比较适用于样本容量比较大的类域的自动分类，而那些样本容量较小的类域采用这种算法比较容易产生误分。

KNN 算法的决策过程如下：

在下图中，圆要被决定赋予哪个类，是三角形还是四方形？如果 k=3，那么由于三角形所占比例为 2/3，所以圆将被赋予三角形那个类；如果 k=5，那么由于四方形比例为

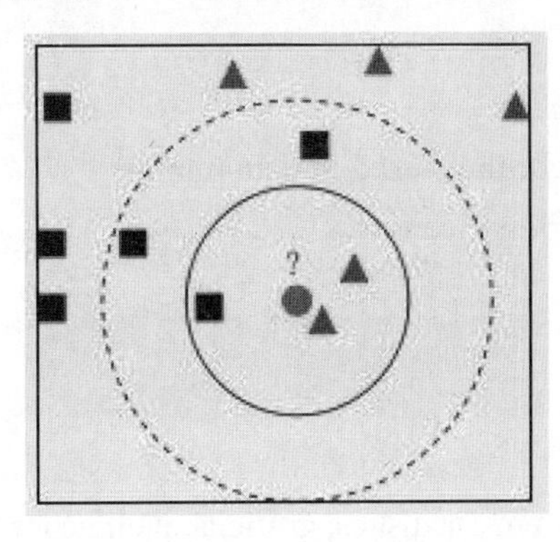

3/5，所以圆被赋予四方形类。

7. Confusion matrix

In the field of machine learning and specifically the problem of statistical classification, a confusion matrix, also known as an error matrix, is a specific table layout that allows visualization of the performance of an algorithm, typically a supervised learning one (in unsupervised learning it is usually called a matching matrix). Each row of the matrix represents the instances in an actual class while each column represents the instances in a predicted class, or vice versa.

混淆矩阵：在人工智能中，混淆矩阵是可视化工具，特别用于监督学习（监督学习是利用一组已知类别的样本调整分类器的参数，使其达到所要求性能的过程，也称为监督训练或有教师学习）。在图像精度评价中，主要用于比较分类结果和实际测得值，可以把分类结果的精度显示在一个混淆矩阵里面。混淆矩阵是通过将每个实测像元的位置和分类与分类图像中的相应位置和分类像比较计算的。

Comprehension

Blank Filling

1. Data mining involves _______ and _________ large amounts of data to find _________ for big data. The techniques came out of the fields of ________ and ________, with a bit of ____________ thrown into the mix.
2. The goal of the data mining is either ___________ or ________. In classification, the idea is to sort data into _______.In prediction, the idea is to predict the value of a _______ variable.
3. Typical algorithms used in data mining include the following: ____________________, ___________, ____________, __________, etc.
4. Logistic regression produces a formula that predicts the ________ of the occurrence as a ________ of the independent variables.
5. The network consists of __________, ___________, and ________. Each unit is assigned a weight. Data is given to the input node, and by a system of ______ and _______, the algorithm adjusts the weights until it meets a certain _____________.

Content Questions

1. What is the data mining?
2. What are the typical algorithms used in data mining?
3. What is the process of classification?

Answers

Blank Filling

1. exploring; analyzing; patterns; statistics; artificial intelligence (AI); database management

2. classification; prediction; groups; continuous
3. Classification trees; Logistic regression; Neural networks; Clustering techniques
4. probability; function
5. input nodes; hidden layers; output nodes; trial; error; stopping criteria

Content Questions

1. Data mining involves exploring and analyzing large amounts of data to find patterns for big data.
2. Classification trees, Logistic regression, Neural networks, K-nearest neighbors.
3. Select the training data, establish classification model, data classification.

参考译文

数据挖掘是指通过分析大量数据来找到特定模式的过程。数据挖掘技术通过结合统计、人工智能、数据管理等诸多方法来实现上述目标。

通常，数据挖掘的目标是分类或预测。分类是将数据分组并排序。例如，营销人员可能会对那些回应促销的人与那些没有回应促销的人的特征感兴趣。

预测与分类不同。在预测中，其核心是预测一个连续变量的值。例如，营销人员可能对预测哪些人会对促销做出反应感兴趣。

数据挖掘中使用的典型算法如下。

分类树：一种流行的数据挖掘技术，用于根据一个或多个预测变量的测量对因变量进行分类。结果是一棵树，节点和节点之间的连接形成 if-then 规则。

可用一棵树来显示泰坦尼克号上乘客生存情况，sibsp 是船上配偶或兄弟姐妹的人数。叶子节点的数字显示了结果的可能性以及其所占比例，如图 2-1 所示。

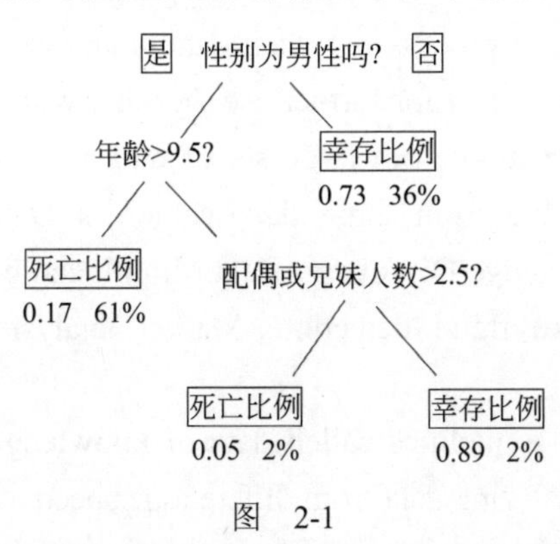

图 2-1

逻辑回归：一种统计学技术，是标准回归的变形，但扩展了处理分类的概念。逻辑回归公式可将发生概率作为自变量的函数进行预测。

神经网络：一种模仿动物神经网络行为特征，进行分布式并行信息处理的算法数学模型。网络由输入节点、隐藏层和输出节点组成。每个节点被分配一个权重。数据连接输入节点，并且通过试错系统调整权重，直到满足一定的停止准则。该方法可以看作一个黑盒方法。

聚类技术［如 K 最近邻（K-Nearest Neighbor，KNN）分类算法］：它是一种识别类似记录组的技术。K 最近邻算法计算历史数据中记录与点之间的距离。然后，它将该记录分配给数据集中最近邻的类。

给出一个分类树示例。考虑一个电话公司想要确定哪些住宅客户可能会断开其服务的情况。电话公司拥有以下属性的信息：用户拥有这项服务多久，用户在服务上花了多少时间，服务是否一直是有问题的，用户是否拥有他所需要的最好的电话通信计划，用户住的地方，用户年龄多大，用户是否有其他服务捆绑在一起，有关其他运营商计划的竞争信息，以及用户是否仍然使用该服务。

当然，可以找到比这更多的属性。最后一个属性是结果变量，软件将客户分为以下两组：忠实用户和有停机可能的用户。

数据集分为训练数据和测试数据。训练数据包括观察值（称为属性）和结果变量（在分类模型的情况下为二进制），结果变量在本例中即为停留可能和停机可能。

该算法运行在训练数据上，并提出了可以像一系列规则一样读取的树。例如，如果用户已经使用公司服务十多年，且他们已经 55 岁以上，那么他们很有可能仍然是忠实用户。

然后将这些规则运行在测试数据上，以确定该模型对“新数据”有多适用。为模型提供了准确性度量。例如，混淆矩阵是目前流行的技术，该矩阵的表格可以比较案例分类结果的正确性。

如果模型看起来不错，它就可以用在其他数据上，因为它是可用的（去掉即使用它来预测停止业务风险的新情况）。根据模式分析，公司可能会决定出台一些政策，如，向那些被模式分析检测出的具有停止业务风险的用户给予特别优惠。

Text B

Data Mining is one important way to analyze the data in some proper format. Data Mining is a process in which data is analyzed on different criteria and summarize it for further use. In other words Data Mining is extract information from large set of data values. That means mining knowledge from large data values is Data Mining also referred as Knowledge Discovery. Data Mining can be useful in different areas like fraud detection, Market analysis, Target Analysis.

Generally, data mining (sometimes called data or knowledge discovery) is the process of analyzing data from different perspectives and summarizing it into useful information — information that can be used to increase revenue, cuts costs, or both. Technically, data mining is the process of finding correlations or patterns among dozens of fields in large relational database. Data mining as a term used for the specific classes of six activities or tasks as follows:

- Classification
- Estimation

New Words and Expressions

estimation/ˌestɪˈmeɪʃn/ *n.* 估计

- Prediction
- Association rules
- Clustering
- Description

A. Classification

Classification is a process of generalizing the data according to different instances. Several major kinds of classification algorithms in data mining are Decision tree, K-Nearest Neighbor classifier, Naive Bayes, Apriori and AdaBoost. Classification consists of examining the features of a newly presented object and assigning to it a predefined class. The classification task is characterized by the well-defined classes, and a training set consisting of reclassified examples.

B. Estimation

Estimation deals with continuously valued outcomes. Given some input data, we use estimation to come up with a value for some unknown continuous variables such as income, height or credit card balance.

C. Prediction

It's a statement about the way things will happen in the future, often but not always based on experience or knowledge. Prediction may be a statement in which some outcome is expected.

D. Association Rules

An association rule is a rule which implies certain association relationships among a set of objects (such as "occur together" or "one implies the other") in a database.

E. Clustering

Clustering can be considered the most important unsupervised learning problem; so, as every other problem of this kind, it deals with finding a structure in a collection of unlabeled data.

Challenges of Big Data Mining

Volume and Scalability

It is the biggest challenge to deal with the size of data. As Twitter generates 7 + Terabytes of data and Facebook generate 10 + Terabytes of data every year so it becomes difficult to manage and analyze. As we are moving from Terabytes to Petabytes and from Petabytes to Zeta bytes of data it's the important task to analyze this Big Data by some methodology. Scale the data in proper way is the important issue in big data mining.

New Words and Expressions

association rules
关联规则

balance/ˈbæləns / *n.*
余额

imply/ɪmˈplaɪ/ *v.*
暗示；意味；隐含；说明，表明

unsupervised/ˌʌnˈsjuːpəvaɪzd/
adj. 无人监督的，无人管理的

methodology/meθəˈdɒlədʒi]/ *n.*
方法论；方法学；（从事某一活动的）一套方法

New Words and Expressions

heterogeneity/ˌhetərəˈdʒəˈniːətɪ/ *n.* 不均匀性

homogeneous/ˌhɒməˈdʒiːniəs/ *adj.* 均匀的；同性质的，同类的

MisHandling of Big Data

Data handling mainly depends on the scalability of data. And scalability depends on data size, hardware size, and concurrency. Day by day data size is increasing and format to store data is also changing and not fixed in future so it's the task of data analyst to overcome such challenge as mishandling of data by different users.

Privacy and Security

In big data, data size and format are not fixed so it's difficult to maintain privacy of one user from another. And because of this volume of data security algorithms are not fixed. When size of data changes or format changes then we need to apply new security algorithms. Ones we define the security or privacy algorithms to it cannot be applicable to upgraded data. E.g., in hospital the data collected and it may upgrade daily and it may be in different format, so it becomes difficult to analyze and secure the newly added data. As data is linked with so many formats and users it's a fear to keep privacy of data and hence it's a big challenge in data mining.

Speed and Velocity

Velocity refers to unique speed with timely manner. But in many cases it is difficult to maintain unique speed because of variety and size of data.

Heterogeneity of Data

Data analysis has first step that data must be structured in a well format. Some errors and confusion in data may lead to miss classification of data. Machine analysis algorithm only understands homogeneous or structured data. Hence to make the data in homogeneous format is a big challenge in big data mining.

Note:

The text is adapted from the website:

http://xueshu.baidu.com/s?wd=paperuri%3A%286aa1d01e1827b6e7759a6bbb7c098144%29&filter=sc_long_sign&sc_ks_para=q%3DBig%20Data%20Mining%3A%20Challenges%2C%20Technologies%2C%20Tools%20and%20Applications&sc_us=9685418709125473720&tn=SE_baiduxueshu_c1gjeupa&ie=utf-8.

参考译文

数据挖掘是以一定格式分析数据的重要方法之一，是一个以不同标准分析数据的过程，并对其进行总结以供人们进一步使用。即数据挖掘是从大量数据值中提取信息。我们

把从大数据中挖掘知识称为数据挖掘，也称为知识发现。数据挖掘可用于不同的领域，如欺诈检测、市场分析、目标分析。

通常，数据挖掘（有时称为数据发现或知识发现）是从不同角度分析数据并将其总结为有用信息的过程，有用信息即可用于增加收入、降低成本或同时实现两者的信息。从技术上讲，数据挖掘是在大型关系数据库中发现数十个领域之间的关联或模式的过程。数据挖掘用于以下六类活动或任务：

分类、估计、预测、关联规则、聚类、说明。

A．分类

分类是根据不同实例推广数据的过程。数据挖掘中主要的分类算法有决策树、K 最近邻分类器、朴素贝叶斯算法、Apriori 算法和 AdaBoost 算法。分类包括检查新提出的对象的特征并为其分配预定义的类。分类任务的特征在于明确定义的类，以及由重分类示例组成的训练集。

B．估计

估计处理连续值的结果。给出一些输入数据，来估计得出一些未知的连续变量的值，例如收入、高度或信用卡余额。

C．预测

这是关于未来发生的事情的描述，并不总是基于经验或知识。预测可能表达了对某种结果的期待。

D．关联规则

关联规则就是有关联的规则，它意味着数据库中一组对象之间的某些关联关系（如“一起出现”或“一个隐含另一个”）。

E．聚类

聚类可以被认为是最重要的无监督学习问题；所以，像类似的其他问题一样，它可以实现在未标记数据的集合中找到一个结构。

大数据挖掘的挑战

数据集和可扩展性

处理大量数据是最大的挑战。目前，Twitter 每年产生 7TB 以上的数据，Facebook 每年都会产生 10TB 以上的数据，因此变得难以管理和分析。当我们的数据规模从太字节（TB）转向拍字节（PB），从拍字节（PB）转向泽字节（ZB）时，通过某种方法分析这个大数据是重要的任务。大数据挖掘中的重要问题是正确地量化数据。

大数据的误操作

数据处理主要取决于数据的可扩展性。可扩展性取决于数据大小、硬件大小和并发性。数据量日益增加，存储数据的格式也在变化，并且在未来也不会固定不变，所以数据分析师的任务是克服诸如对不同用户的数据处理不当的挑战。

隐私和安全

在大数据中，数据大小和格式不是固定的，很难在用户之间保护隐私。而且由于这个数据量的数据安全算法并不固定，当数据大小更改或格式更改时，我们需要应用新的安全算法。我们定义安全性或隐私算法，不能适用于数据升级。例如，医院收集的数据可能每天升级，可能会有不同的格式，因此难以分析和保护新增的数据。由于数据与许多格式和用户相关联，因此担心保护数据的隐私，这在数据挖掘中是一个很大的挑战。

速度和高速

高速是指随时间变化的具有一定模式的速度。但是在许多情况下，由于数据的种类和大小，难以保持某种特定模式的速度。

异构数据

数据分析的第一步是数据必须以良好的格式进行结构化。数据中的一些错误和混淆可能导致数据错误分类。机器分析算法只能理解均匀或结构化数据。因此，使数据格式统一成为大数据挖掘的一大挑战。

Chapter 3

Big Data Analytics

Text A

Big data analytics is the process of examining large data sets to uncover hidden patterns, unknown correlations, market trends, customer preferences and other useful business information. The analytical findings can lead to more effective marketing, new revenue opportunities, better customer service, improved operational efficiency, competitive advantages over rival organizations and other business benefits.

The primary goal of big data analytics is to help companies make more informed business decisions by enabling data scientists, predictive modelers and other analytics professionals to analyze large volumes of transaction data[1], as well as other forms of data that may be untapped by conventional business intelligence(BI) programs. That could include Web server logs and Internet clickstream data, social media content and social network activity reports, text from customer emails and survey responses, mobile-phone call detail records and machine data captured by sensors connected to the Internet of Things.

Semi-structured and unstructured data may not fit well in traditional data warehouses based on relational databases[2]. Furthermore, data warehouses may not be able to handle the processing demands posed by sets of big data that need to be updated frequently or even continually — for example, real-time data on the performance of mobile applications or of oil and gas

New Words and Expressions

rival/ˈraɪvəl/ *n.*
竞争对手；敌手

predictive/prɪˈdɪktɪv/ *adj.*
预言性的，预测的；前瞻的

informed/ɪnˈfɔːmd/ *adj.*
了解情况的；见多识广的；消息灵通的

transaction data
事务数据，事务处理数据，交易数据

untapped/ʌnˈtæpt/ *adj.*
未利用的；未开发的

clickstream
【计】点击流（指对网络用户上网点击的一系列网页的记录）

Internet of Things
物联网

relational database
关系数据库，关系型数据库，关联式资料库

pipelines. As a result, many organizations looking to collect, process and analyze big data have turned to a newer class of technologies that includes Hadoop[3] and related tools such as YARN, MapReduce[4], Spark[5], Hive[6] and Pig[7] as well as NoSQL database[8]. Those technologies form the core of an open source software framework that supports the processing of large and diverse data sets across clustered systems.

In some cases, Hadoop clusters and NoSQL systems are being used as landing pads and staging areas for data before it gets loaded into a data warehouse for analysis, often in a summarized form that is more conducive to relational structures. Increasingly though, big data vendors are pushing the concept of a Hadoop data lake that serves as the central repository for an organization's incoming streams of raw data. In such architectures, subsets of the data can then be filtered for analysis in data warehouses and analytical databases, or it can be analyzed directly in Hadoop using batch query tools, stream processing software and SQL[9] on Hadoop technologies that run interactive, ad hoc queries[10] written in SQL.

Big data can be analyzed with the software tools commonly used as part of advanced analytics disciplines such as predictive analytics, data mining, text analytics and statistical analysis. Mainstream BI software and data visualization tools can also play a role in the analysis process.

Potential pitfalls that can trip up organizations on big data analytics initiatives include a lack of internal analytics skills and the high cost of hiring experienced analytics professionals. The amount of information that's typically involved, and its variety, can also cause data management headaches, including data quality and consistency issues. In addition, integrating Hadoop systems and data warehouses can be a challenge, although various vendors now offer software connectors between Hadoop and relational databases, as well as other data integration tools with big data capabilities.

Why is big data analytics important?

Big data analytics helps organizations harness their data and use it to identify new opportunities. That, in turn, leads to smarter business moves, more efficient operations, higher profits and happier customers. In his report Big Data in Big Companies, IIA Director of Research Tom Davenport interviewed more than 50

New Words and Expressions

Hadoop
分布式计算

diverse/daɪˈvɜːs/ *adj.*
不同的，多种多样的；形形色色的

clustered system
【计】群集系统

sensor/ˈsensə(r)/ *n.*
传感器

transmit/trænsˈmɪt/ *v.*
播送，发射，传送（信号）

velocity/vəˈlɒsəti/ *n.*
速度；速率

extract/ɪkˈstrækt/ *v.*
提取

optimal/ˈɒp.tɪ.məl/ *adj.*
最优的，最佳的；优化的

analytics/ˌænəˈlɪtɪks/ *n.*
分析，逻辑分析的方法

exceed/ɪkˈsiːd/ *v.*
超过；胜过

conducive/kənˈdʒuː sɪv/ *adj.*
有利的，有助的，有益的

vendor/ˈvendər/ *n.*
卖主

repository/rɪˈpɒzɪtəri/ *n.*
仓库；贮藏室；存放处

raw data
原始数据

subset/ˈsʌbset/ *n.*
子集（类似的数字、物体或人员的一个集合组，是另一个较大集合的一部分）

batch/bætʃ/ *n.*
一批；一批生产的量

interactive /ˌɪntəˈræktɪv/ *adj.*
互相作用的；【计】交互式的

pitfall/ˈpɪtfɔːl/ *n.*
隐患；陷阱

businesses to understand how they used big data. He found they got value in the following ways:

(1) Cost reduction. Big data technologies such as Hadoop and cloud-based analytics[11] bring significant cost advantages when it comes to storing large amounts of data — plus they can identify more efficient ways of doing business.

(2) Faster, better decision making. With the speed of Hadoop and in-memory analytics[12], combined with the ability to analyze new sources of data, businesses are able to analyze information immediately — and make decisions based on what they've learned.

(3) New products and services. With the ability to gauge customer needs and satisfaction through analytics comes the power to give customers what they want. Davenport points out that with big data analytics, more companies are creating new products to meet customers' needs.

New Words and Expressions

trip up
绊，绊倒（某人）

consistency/kənˈsɪstənsi/ *n.*
连贯性；一致性

harness/ˈhɑːnɪs/ *n./v.*
背带；利用，控制

interview /ˈɪntəvjuː/ *n./v.*
接见，采访；面试；访问

gauge /ɡeɪdʒ/ *n./v.*
尺度，标准；测量，评估；采用

cost reduction
降低成本

Note:

The text is adapted from the website:

http://searchbusinessanalytics.techtarget.com/definition/big-data-analytics.

Terms

1. Transaction data

Transaction data are data describing an event (the change as a result of a transaction) and is usually described with verbs. Transaction data always has a time dimension, a numerical value and refers to one or more objects (i.e. the reference data).

Typical transactions are:

- Financial — orders, invoices, payments
- Work — plans, activity records
- Logistics — deliveries, storage records, travel records, etc.

Typical transaction processing systems (systems generating transactions) are SAP and Oracle Financials.

事务数据是描述事件（作为事务结果的更改）的数据，并且通常用动词来描述。事务数据总是具有时间维度和数值，并且指代一个或多个对象（即参考数据）。

典型交易包括：

- 财务——订单、发票、付款；
- 工作——计划、活动记录；
- 物流——交货、仓储记录、旅行记录等。

典型的交易处理系统（生成交易的系统）是 SAP 和 Oracle Financials。

2. Relational database

A relational database is a digital database whose organization is based on the relational model of data, as proposed by E. F. Codd in 1970. The various software systems used to maintain relational databases are known as a relational database management system (RDBMS). Virtually all relational database systems use SQL (Structured Query Language) as the language for querying and maintaining the database.

关系数据库是数字数据库，其组织方式是基于 E. F. Codd 于 1970 年提出的关系数据模型。用于维护关系数据库的各种软件系统被称为关系数据库管理系统（RDBMS）。几乎所有关系数据库系统都使用 SQL（结构化查询语言）作为查询和维护数据库的语言。

3. Hadoop

Apache Hadoop (/hə'du：p/) is an open-source software framework used for distributed storage and processing of big data sets using the MapReduce programming model. It consists of computer clusters built from commodity hardware. All the modules in Hadoop are designed with a fundamental assumption that hardware failures are common occurrences and should be automatically handled by the framework.

The core of Apache Hadoop consists of a storage part, known as Hadoop Distributed File System (HDFS), and a processing part which is a MapReduce programming model. Hadoop splits files into large blocks and distributes them across nodes in a cluster. It then transfers packaged code into nodes to process the data in parallel. This approach takes advantage of data locality — nodes manipulating the data they have access to — to allow the dataset to be processed faster and more efficiently than it would be in a more conventional supercomputer architecture that relies on a parallel file system where computation and data are distributed via high-speed networking.

The base Apache Hadoop framework is composed of the following modules:

- Hadoop Common — contains libraries and utilities needed by other Hadoop modules;
- Hadoop Distributed File System (HDFS) — a distributed file-system that stores data on commodity machines, providing very high aggregate bandwidth across the cluster;
- Hadoop YARN — a resource-management platform responsible for managing computing resources in clusters and using them for scheduling of users' applications;
- HadoopMapReduce — an implementation of the MapReduce programming model for large scale data processing.

The term Hadoop has come to refer not just to the base modules above, but also to the ecosystem, or collection of additional software packages that can be installed on top of or alongside Hadoop, such as Apache Pig, Apache Hive, Apache HBase, Apache Phoenix, Apache Spark, Apache ZooKeeper, Cloudera Impala, Apache Flume, Apache Sqoop, Apache Oozie, Apache Storm.

Apache Hadoop 是使用 MapReduce 编程模型开发分布式存储和处理大数据集的开源软件框架。它是由商品硬件构成的计算机集群。Hadoop 中的所有模块都设计了一个基本假设，即硬件故障是常见的情况，应由框架自动处理。

Apache Hadoop 的核心包括称为 Hadoop 分布式文件系统（Hadoop Distributed File System，HDFS）的存储部分，以及 MapReduce 编程模型的处理部分。Hadoop 将文件分解成大块，并将它们分布在集群中的节点上。然后将打包的代码传输到节点中并行处理。这种方法利用了数据的局部性——操纵他们访问的数据的节点——使数据集的处理速度和效率比依赖于一个并行文件系统的传统超级计算机体系结构要快得多、高效得多，传统方式中计算和数据是通过高速网络分发的。

基础 Apache Hadoop 框架由以下模块组成：

Hadoop Common——包含其他 Hadoop 模块所需的库和实用程序；

Hadoop 分布式文件系统（HDFS）—— 一种分布式文件系统，用于存储商品机上的数据，在集群中提供非常高的聚合带宽；

Hadoop YARN——一个资源管理平台，负责管理集群中的计算资源，并将其用于调度用户应用程序；

HadoopMapReduce——用于大规模数据处理的 MapReduce 编程模型的实现。

Hadoop 这一术语不仅仅指上面的基本模块，还涉及一套系统，或者说可以安装在 Hadoop 之上或之外的其他软件包的集合，例如 Apache Pig，Apache Hive，Apache HBase，Apache Phoenix Apache Spark，Apache ZooKeeper，Cloudera Impala，Apache Flume，Apache Sqoop，Apache Oozie，Apache Storm。

4. MapReduce

MapReduce is a programming model and an associated implementation for processing and generating big data sets with a parallel, distributed algorithm on a cluster.

A MapReduce program is composed of a Map procedure (method) that performs filtering and sorting (such as sorting students by first name into queues, one queue for each name) and a Reduce method that performs a summary operation (such as counting the number of students in each queue, yielding name frequencies). The "MapReduce System" (also called "infrastructure" or "framework") orchestrates the processing by marshalling the distributed servers, running the various tasks in parallel, managing all communications and data transfers between the various parts of the system, and providing for redundancy and fault tolerance.

MapReduce 是一种编程模型，用于在集群上使用并行分布式算法处理和生成大数据集的相关实现。

MapReduce 程序由执行过滤和排序的映射过程（方法）组成（例如将学生按姓氏排列成队列，每个姓氏排一列）以及执行汇总操作的 Reduce 方法（例如，数出每个队列中的学生数量，得到姓氏频率）。“MapReduce 系统”（也称为“基础架构”或“框架”）通过编组分布式服务器精心安排进程，并行运行各种任务，管理系统各部分之间的所有通信和数据传输，并提供冗余和容错能力。

5. Spark

Apache Spark, a cluster computing framework. Apache Spark has an advanced DAG execution engine that supports acyclic data flow and in-memory computing. Spark offers over 80 high-level operators that make it easy to build parallel apps. And you can use it interactively from the Scala, Python and R shells. Spark powers a stack of libraries including SQL and Data Frames, MLlib for machine learning, GraphX, and Spark Streaming. You can combine these libraries seamlessly in the same application. Spark runs on Hadoop, Mesos, standalone, or in the cloud. It can access diverse data sources including HDFS, Cassandra, HBase, and S3.

Apache Spark 是一个集群计算框架。Apache Spark 具有支持非循环数据流和内存计算的高级 DAG 执行引擎。Spark 提供了超过 80 个高级操作，可以轻松构建并行应用程序。用户可以在 Scala，Python 和 R shell 中交互使用。Spark 支持一系列库，包括 SQL 和 Data Frames，用于机器学习的 MLlib、GraphX 和 Spark Streaming。用户可以在同一应用程序中无缝地组合这些库。Spark 可以独立运行或在 Hadoop、Mesos、云端运行。它可以访问不同的数据源，包括 HDFS、Cassandra、HBase 和 S3。

6. Hive

Apache Hive is a data warehouse infrastructure built on top of Hadoop for providing data summarization, query, and analysis. Hive gives an SQL-like interface to query data stored in various databases and file systems that integrate with Hadoop. Traditional SQL queries must be implemented in the MapReduce Java API to execute SQL applications and queries over distributed data. Hive provides the necessary SQL abstraction to integrate SQL-like Queries (HiveQL) into the underlying Java API without the need to implement queries in the low-level Java API. Since most data warehousing applications work with SQL-based querying languages, Hive supports easy portability of SQL-based application to Hadoop. While initially developed by Facebook, Apache Hive is now used and developed by other companies such as Netflix and the Financial Industry Regulatory Authority(FINRA). Amazon maintains a software fork of Apache Hive that is included in Amazon Elastic MapReduce on Amazon Web Services.

Apache Hive 是建立在 Hadoop 之上的数据仓库基础架构，用于提供数据汇总、查询和分析。Hive 提供了类似 SQL 的界面来查询存储在与 Hadoop 集成的各种数据库和文件系统中的数据。传统的 SQL 查询必须在 MapReduce Java API 中实现，以便对分布式数据执行 SQL 应用程序和查询。Hive 提供必要的 SQL 抽象，以将 SQL 类似的查询（HiveQL）集成到底层 Java API 中，而无须在低级 Java API 中实现查询。由于大多数数据仓库应用程序都支持基于 SQL 的查询语言，所以 Hive 支持将基于 SQL 的应用程序很容易地移植到 Hadoop。虽然最初由 Facebook 开发，Apache Hive 现在由其他公司［如 Netflix 和金融业监管机构（FINRA）］使用和开发。亚马逊维护包含在其网络服务 Amazon Elastic MapReduce 中的 Apache Hive 软件分支。

7. Pig

Pig (programming tool) is a high-level platform for creating programs that run on Apache Hadoop. The language for this platform is called Pig Latin. Pig can execute its Hadoop jobs in MapReduce, Apache Tez, or Apache Spark. Pig Latin abstracts the programming from the Java MapReduce idiom into a notation which makes MapReduce programming high level, similar to that of SQL for RDBMSs. Pig Latin can be extended using User Defined Functions (UDFs) which the user can write in Java, Python, JavaScript, Ruby or Groovy and then call directly from the language.

Pig（编程工具）是在 Apache Hadoop 上运行的创建程序的高级平台。这个平台的语言叫作 Pig Latin。Pig 可以在 MapReduce、Apache Tez 或 Apache Spark 中执行其 Hadoop 工作。使用 Pig 进行数据处理、分析时，需要使用其提供的 Pig Latin 脚本语言编写相应脚本，这些脚本执行时会被转换成 Map 和 Reduce 任务，与 RDBMS 的 SQL 类似。Pig Latin 可以使用用户定义的函数（UDF）进行扩展，这些函数可以使用 Java、Python、JavaScript、Ruby

或 Groovy 编写，然后直接调用。

8. NoSQL database

A NoSQL (originally referring to “non SQL”, “non relational” or “not only SQL”) database provides a mechanism for storage and retrieval of data which is modeled in means other than the tabular relations used in relational databases. Such databases have existed since the late 1960s, but did not obtain the “NoSQL” moniker until a surge of popularity in the early twenty-first century, triggered by the needs of Web 2.0 companies such as Facebook, Google, and Amazon.com. NoSQL databases are increasingly used in big data and real-time web applications. NoSQL systems are also sometimes called “Not only SQL” to emphasize that they may support SQL-like query languages.

Motivations for this approach include: simplicity of design, simpler “horizontal” scaling to clusters of machines (which is a problem for relational databases), and finer control over availability. The data structures used by NoSQL databases (e.g. key-value, wide column, graph, or document) are different from those used by default in relational databases, making some operations faster in NoSQL. The particular suitability of a given NoSQL database depends on the problem it must solve. Sometimes the data structures used by NoSQL databases are also viewed as “more flexible” than relational database tables.

NoSQL 数据库（最初指的是“非 SQL”“非关系”或“不只是 SQL”）提供了一种用于存储和检索数据的机制，该机制以关系数据库中使用的表格关系以外的方式建模。这些数据库自 20 世纪 60 年代后期就已经存在，直到 21 世纪初，由于 Facebook、Google 和 Amazon.com 这样的 Web 2.0 公司的需求引发了人气的激增，它们才获得 NoSQL 的绰号。NoSQL 数据库越来越多地用于大数据和实时网络应用程序。NoSQL 系统有时也被称为“不仅仅是 SQL”，强调他们可以支持类似 SQL 的查询语言。

这种方法的动机包括：简单的设计、更简单的“横向”缩放到机器集群（这是关系数据库的一个问题）以及更好地控制可用性。NoSQL 数据库使用的数据结构（例如键值、宽列、图形或文档）与关系数据库中默认使用的数据结构不同，这使得 NoSQL 中的某些操作更快。给定的 NoSQL 数据库的特殊适用性取决于它必须解决的问题。有时，NoSQL 数据库使用的数据结构也被视为比关系数据库表更“灵活”。

9. SQL

SQL (Structured Query Language) is a domain-specific language used in programming and designed for managing data held in a relational database management system (RDBMS), or for stream processing in a relational data stream management system (RDSMS).

Originally based upon relational algebra and tuple relational calculus, SQL consists of a data definition language, data manipulation language, and data control language. The scope of SQL includes data insert, query, update and delete, schema creation and modification, and data access control. Although SQL is often described as, and to a great extent is, a declarative language (4GL), it also includes procedural elements.

SQL was one of the first commercial languages for Edgar F. Codd’s relational model, as described in his influential 1970 paper, “A Relational Model of Data for Large Shared Data

Banks." Despite not entirely adhering to the relational model as described by Codd, it became the most widely used database language.

SQL（结构化查询语言）是一种特定用于数据领域的语言，用于编程和设计管理关系数据库管理系统（RDBMS）中保存的数据，或用于关系数据流管理系统（RDSMS）中的流处理。

SQL 最初是基于关系代数和元组关系演算，由数据定义语言、数据操纵语言和数据控制语言组成。SQL 的范围包括数据插入、查询、更新、删除、模式创建和修改以及数据访问控制。尽管 SQL 通常被描述为并且在很大程度上是声明性语言（4GL），它也包含过程元素。

如在著名的 1970 年的文章"大型共享数据库数据的关系模型"中所述，SQL 是 Edgar F. Codd 的关系模型的第一种商业语言之一。尽管没有完全遵循 Codd 所述的关系模型，但它已成为使用最为广泛的数据库语言。

10. Ad hoc queries

Ad hoc queries are single questions or requests for a database written in SQL or another query language by the user on-demand — typically when the user needs information outside of regular reporting or predefined queries.

特定数据查询是用户根据自己的需求灵活地选择查询条件，系统能够根据用户的选择生成相应的统计报表。特定数据查询与普通应用查询最大的不同是普通的应用查询是定制开发的，而特定数据查询是由用户自定义查询条件的。

11. Cloud-based analytics

Cloud-based analytics is a marketing term for businesses to carry out analysis using cloud computing. It uses a range of analytical tools and techniques to help companies extract information from massive data and present it in a way that is easily categorized and readily available via a web browser.

云分析是企业使用云计算进行分析的营销术语。它使用了一系列分析工具和技术，帮助企业从海量数据中提取信息，并以一种易于分类和易于通过网络浏览器获得的方式呈现信息。

12. In-memory analytics

In-memory analytics is an approach to querying data when it resides in a computer's random access memory (RAM), as opposed to querying data that is stored on physical disks. This results in vastly shortened query response times, allowing business intelligence (BI) and analytic applications to support faster business decisions.

As the cost of RAM declines, in-memory analytics is becoming feasible for many businesses. BI and analytic applications have long supported caching data in RAM, but older 32-bit operating systems provided only 4GB of addressable memory. Newer 64-bit operating systems, with up to 1 terabyte (TB) addressable memory (and perhaps more in the future), have made it possible to cache large volumes of data — potentially an entire data warehouse or data mart — in a computer's RAM.

In addition to providing incredibly fast query response times, in-memory analytics can reduce or eliminate the need for data indexing and storing pre-aggregated data in OLAP cubes or aggregate tables. This reduces IT costs and allows faster implementation of BI and analytic applications. It is anticipated that as BI and analytic applications embrace in-memory analytics, traditional data warehouses may eventually be used only for data that is not queried frequently.

内存分析是在数据驻留于计算机的随机存取存储器（RAM）中时查询数据的方法，而不是查询存储在物理磁盘上的数据。这就大大缩短了查询响应时间，从而允许商业智能(BI)和分析应用程序支持更快的业务决策。

随着 RAM 成本的下降，内存分析对于许多企业而言变得可行。商业智能和分析应用程序长期支持在 RAM 中缓存数据，但较旧的 32 位操作系统只提供 4GB 的可寻址内存。较新的 64 位操作系统具有高达 1TB 的可寻址内存（将来可能会更大），使得在计算机的 RAM 中缓存整个数据仓库或数据库的大量数据变得可能。

除了提供令人难以置信的快速查询响应之外，内存分析可以减少或消除在 OLAP 数据集或聚合表中进行数据索引并存储预聚合数据的需求。这可以降低 IT 成本，并可以使商业智能和分析应用程序更快地实施。由于商业智能和分析应用程序在内存分析中的应用，传统的数据仓库最终只能用于存储不经常查询的数据。

Comprehension

Blank Filling

1. Big data analytics is the process of examining large data sets to uncover hidden _______, unknown _________, market _______, customer _________ and other useful business information.
2. The primary goal of big data analytics is to help companies make more informed business decisions by enabling ____________, predictive _________ and other analytics professionals to analyze large volumes of ______________, as well as other forms of data that may be untapped by conventional business intelligence(BI) programs.
3. Many organizations looking to ________, ________ and ________ big data have turned to a newer class of technologies that includes Hadoop and related ______ such as YARN, MapReduce, Spark, Hive and Pig as well as NoSQL database. Those technologies form the core of an open source software ________ that supports the processing of large and diverse data sets across ________ systems.
4. In some cases, Hadoop clusters and NoSQL systems are being used as ____________ and ____________ for data before it gets loaded into a data warehouse for analysis, often in a summarized form that is more ____________ to relational structures.
5. Big data can be analyzed with the software tools commonly used as part of advanced analytics disciplines such as ___________, ____________, ____________ and ____________.
6. Big data analytics helps organizations harness their data and use it to ______________. That, in turn, leads to smarter ____________, more efficient operations, higher _______ and happier customers.

Content Questions

1. What is the big data analytics?

2. What can analytical finding of big data lead to?
3. What is the primary goal of data analytics?
4. What kind of data is analyzed by data analytics?
5. Why do organization turn to big data analytics?
6. What technologies form the core of an open source software framework?
7. What are the potential pitfalls that can trip up organizations on big data analytics initiatives?
8. In what ways do businesses get values from big data analytics?

Answers

Blank Filling

1. patterns; correlations; trends; preferences
2. data scientists; modelers; transaction data
3. collect; process; analyze; tools; framework; clustered
4. landing pads; staging areas; conducive
5. predictive analytics; data mining; text analytics; statistical analysis
6. identify new opportunities; business moves; profits

Content Questions

1. Big data analytics is the process of examining large data sets to uncover hidden patterns, unknown correlations, market trends, customer preferences and other useful business information.
2. The analytical findings can lead to more effective marketing, new revenue opportunities, better customer service, improved operational efficiency, competitive advantages over rival organizations and other business benefits.
3. The primary goal of big data analytics is to help companies make more informed business decisions by enabling data scientists, predictive modelers and other analytics professionals to analyze large volumes of transaction data1, as well as other forms of data that may be untapped by conventional business intelligence(BI) programs.
4. Analyze large volumes of transaction data, as well as other forms of data that may be untapped by conventional business intelligence(BI) programs. That could include Web server logs and Internet clickstream data, social media content and social network activity reports, text from customer emails and survey responses, mobile-phone call detail records and machine data captured by sensors connected to the Internet of Things.
5. Semi-structured and unstructured data may not fit well in traditional data warehouses based on relational databases. Furthermore, data warehouses may not be able to handle the processing demands posed by sets of big data that need to be updated frequently or even continually — for example, real-time data on the performance of mobile applications or of oil and gas pipelines. As a result, many organizations turn to big data analytics.

6. Technologies include Hadoop and related tools such as YARN, MapReduce, Spark, Hive and Pig as well as NoSQL database.
7. Potential pitfalls that can trip up organizations on big data analytics initiatives include a lack of internal analytics skills and the high cost of hiring experienced analytics professionals.
8. They got value in the following ways: Cost reduction; Faster, better decision making; New products and services.

参考译文

大数据分析是对大数据集进行检查以发现隐藏模式、未知相关性、市场趋势、客户喜好和其他有用商业信息的过程。分析结果可以带来更有效的营销、新的收入机会、更好的客户服务、提高的运营效率、超过对手的竞争优势和其他业务收益。

大数据分析的主要目标是帮助公司通过数据科学家、预测建模者和其他分析专业人员对大量的交易数据的分析而做出更明智的业务决策，以及可能未被传统商业智能（BI）计划开发出的其他形式的数据。那些数据包括 Web 服务器日志和互联网点击流数据、社交媒体内容和社交网络活动报告、来自客户电子邮件和调查回复的文本、移动电话通话记录和连接到物联网的传感器获得的机器数据。

半结构化和非结构化数据可能不适合基于关系数据库的传统数据仓库。此外，数据仓库可能无法处理需要频繁更新甚至持续更新的大型数据集合所产生的处理需求，例如关于移动应用程序或油气管道性能的实时数据。因此，许多寻求收集、处理和分析大数据的组织已经转向包括 Hadoop 和 YARN、MapReduce、Spark、Hive、Pig 以及 NoSQL 数据库等相关工具在内的新一类技术。这些技术构成了一个开源软件框架的核心，该框架支持跨群集系统处理大型和多样化的数据集。

在某些情况下，在将数据加载到数据仓库进行分析之前 Hadoop 集群和 NoSQL 系统正被用作数据的着陆点和分段区域，通常是更有利于关系结构的概述形式。越来越多的大数据供应商正在推动一个 Hadoop 数据池的概念，该数据池作为组织传入的原始数据流的中央存储库。在这样的架构中，数据的子集可以过滤以便在数据仓库和分析数据库中进行分析，或者可以直接使用 Hadoop 中的批处理查询工具、流处理软件和基于 Hadoop 技术的交互式、特定数据查询的结构化查询语言进行分析。

大数据可以用常用的高级分析软件工具进行分析，例如预测分析、数据挖掘、文本分析和统计分析。主流商业智能软件和数据可视化工具也可以在分析过程中发挥作用。

潜在的缺陷可能使组织在大数据分析上出错，包括缺乏内部分析技能以及雇用具有丰富经验的分析专家的高成本。通常涉及的信息量及其多样性也可能导致数据管理问题，包括数据质量和一致性问题。此外，尽管现在供应商都提供 Hadoop 和关系数据库之间的软件连接，以及具有大数据能力的其他数据集成工具，集成 Hadoop 系统和数据仓库可能仍然存在挑战。

为什么大数据分析很重要？

大数据分析可协助企业挖掘数据，发掘新机遇。这反过来又会带来更智能的商业运作、更高效的运营、更高的利润和更高的客户满意度。在大公司的大数据报告中，IIA 研究总监汤姆·达文波特（Tom Davenport）采访了 50 多家企业，了解他们如何使用大数据。他发现大数据的价值体现在以下几个方面：

（1）降低成本。大数据技术，如 Hadoop 和基于云的分析在存储大量数据方面带来了显著的成本优势，此外，大数据还有助于识别更有效的经商方式。

（2）更快、更好的决策。随着 Hadoop 和内存分析的速度加上分析新数据源的能力，企业能够立即分析信息，并根据所了解到的信息做出决策。

（3）新产品和服务。通过分析来衡量客户需求和满意度的能力使得公司能够为客户提供他们想要的。达文波特指出，通过大数据分析，更多的公司正在创造新产品以满足客户的需求。

Text B

Big Data Analytics with Hadoop

Apache Hadoop was born out of a need to process an avalanche of big data. The web was generating more and more information on a daily basis, and it was becoming very difficult to index over one billion pages of content. In order to cope, Google invented a new style of data processing known as MapReduce. A year after Google published a white paper describing the MapReduce framework, Doug Cutting and Mike Cafarella, inspired by the white paper, created Hadoop to apply these concepts to an open-source software[1] framework to support distribution for the Nutch[2] search engine project. Given the original case, Hadoop was designed with a simple write-once storage infrastructure.

Hadoop has moved far beyond its beginnings in web indexing and is now used in many industries for a huge variety of tasks that all share the common theme of lots of variety, volume and velocity of data — both structured and unstructured. It is now widely used across industries, including finance, media and entertainment, government, healthcare, information services, retail, and other industries with big data requirements but the limitations of the original storage infrastructure remain.

Hadoop is increasingly becoming the go-to framework for large-scale, data-intensive deployments. Hadoop is built to process large amounts of data from terabytes to petabytes and beyond. With this much data, it's unlikely that it would fit on a single computer's hard drive, much less in memory. The beauty of Hadoop is that it is

New Words and Expressions

avalanche/ˈævəlɑːnʃ/ *n.*
雪崩

retail /ˈriːteɪl/ *n./v.*
零售，零卖；转述；传播

terabyte (TB)，petabyte (PB
信息度量单位，1PB=1024TB

designed to efficiently process huge amounts of data by connecting many commodity computers together to work in parallel. Using the MapReduce model, Hadoop can take a query over a dataset, divide it, and run it in parallel over multiple nodes. Distributing the computation solves the problem of having data that's too large to fit onto a single machine.

Hadoop Software

The Hadoop software stack introduces entirely new economics for storing and processing data at scale. It allows organizations unparalleled flexibility in how they're able to leverage data of all shapes and sizes to uncover insights about their business. Users can now deploy the complete hardware and software stack including the OS and Hadoop software across the entire cluster and manage the full cluster through a single management interface.

Apache Hadoop includes a Distributed File System (DFS)[3], which breaks up input data and stores data on the compute nodes. This makes it possible for data to be processed in parallel using all of the machines in the cluster. The Apache Hadoop Distributed File System is written in Java and runs on different operating systems.

Hadoop was designed from the beginning to accommodate multiple file system implementations and there are a number available. DFS and the S3 file system are probably the most widely used, but many others are available, including the MapR File System.

How is Hadoop Different from Past Techniques?

Hadoop can handle data in a very fluid way. Hadoop is more than just a faster, cheaper database and analytics tool. Unlike databases, Hadoop doesn't insist that you structure your data. Data may be unstructured and schemaless. Users can dump their data into the framework without needing to reformat it. By contrast, relational databases require that data be structured and schemas be defined before storing the data.

Hadoop has a simplified programming model. Hadoop's simplified programming model allows users to quickly write and test software in distributed systems. Performing computation on large volumes of data has been done before, usually in a distributed setting but writing software for distributed systems is notoriously hard. By trading away some programming flexibility, Hadoop

New Words and Expressions

commodity/kəˈmɒdəti/ *n.*
商品；有价值的物品

unparalleled /ʌnˈpærəleld/ *adj.*
无比的，无双的，空前的

leverage /ˈliːvərɪdʒ/ *n.*
杠杆作用；优势，力量；影响力

Distributed File System (DFS)
分布式文件系统

fluid /ˈfluːɪd/ *n./adj.*
液体；流动的；易变的，不固定的

dump /dʌmp/ *n./v.*
倾倒；丢下，卸下；摆脱，扔弃；卸货；垃圾场；仓库

relational database
关系数据库

notoriously /nəˈtɔːriəs/ *adv.*
著名地；众所周知地；声名狼藉地

makes it much easier to write distributed programs.

Because Hadoop accepts practically any kind of data, it stores information in far more diverse formats than what is typically found in the tidy rows and columns of a traditional database. Some good examples are machine-generated data and log data, written out in storage formats including JSON, Avro and ORC.

The majority of data preparation work in Hadoop is currently being done by writing code in scripting languages like Hive, Pig or Python.

Hadoop is easy to administer. Alternative High Performance Computing (HPC) systems[4] allow programs to run on large collections of computers, but they typically require rigid program configuration and generally require that data be stored on a separate Storage Area Network (SAN)[5] system. Schedulers on HPC clusters require careful administration and since program execution is sensitive to node failure, administration of a Hadoop cluster is much easier.

Hadoop invisibly handles job control issues such as node failure. If a node fails, Hadoop makes sure the computations are run on other nodes and that data stored on that node are recovered from other nodes.

Hadoop is agile. Relational databases are good at storing and processing data sets with predefined and rigid data models. For unstructured data, relational databases lack the agility and scalability that is needed. Apache Hadoop makes it possible to cheaply process and analyze huge amounts of both structured and unstructured data together, and to process data without defining all structure ahead of time.

Why use Apache Hadoop?

It's cost effective. Apache Hadoop controls costs by storing data more affordably per terabyte than other platforms. Instead of thousands to tens of thousands per terabyte, Hadoop delivers compute and storage for hundreds of dollars per terabyte.

It's fault-tolerant. Fault tolerance is one of the most important advantages of using Hadoop. Even if individual nodes experience high rates of failure when running jobs on a large cluster, data is replicated across a cluster so that it can be recovered easily in the face of disk, node or rack failures.

New Words and Expressions

scripting language
脚本语言

High Performance Computing System
高性能计算机（HPC）系统

configuration /kənˌfɪgəˈreɪʃən/ *n.*
布局，构造；配置；[物]位形，组态

storage area network (SAN) system
独立存储区域网络系统

agile /ˈædʒaɪl/ *adj.*
灵巧的；轻快的；机敏的

scalability/ˌskeɪləˈbɪlə.ti/ *n.*
可量测性

fault tolerance
容错性

replicate/ˈreplɪkeɪt/ *v.*
复制；重复，反复

It's flexible. The flexible way that data is stored in Apache Hadoop is one of its biggest assets — enabling businesses to generate value from data that was previously considered too expensive to be stored and processed in traditional databases. With Hadoop, you can use all types of data, both structured and unstructured, to extract more meaningful business insights from more of your data.

It's scalable. Hadoop is a highly scalable storage platform, because it can store and distribute very large data sets across clusters of hundreds of inexpensive servers operating in parallel. The problem with traditional Relational DataBase Management Systems (RDBMS)[6] is that they can't scale to process massive volumes of data.

New Words and Expressions

flexible/ˈfleksəbəl/ *adj.*
灵活的；易弯曲的；易被说服的

scalable/ˈskeɪləbəl/ *adj.*
可升级的；可扩展的；可攀登的

Relational DataBase Management Systems (RDBMS)
关系数据库管理系统

Note:

The text is adapted from the website:
https://www.mapr.com/products/apache-hadoop.

Terms

1. Open-Source Software

Open-Source Software (OSS) is computer software with its source code made available with a license in which the copyright holder provides the rights to study, change, and distribute the software to anyone and for any purpose. Open-Source Software may be developed in a collaborative public manner. According to scientists who studied it, Open-Source Software is a prominent example of open collaboration. A 2008 report by the Standish Group states that adoption of Open-Source Software models has resulted in savings of about $60 billion (£48 billion) per year to consumers.

In particular, the heightened value proposition from open source in the following categories:

- Security
- Affordability
- Transparency
- Perpetuity
- Interoperability
- Flexibility
- Localization

开源软件（OSS）是提供其源代码许可证的计算机软件，版权所有者可以向任何人以任何目的提供研究、更改和分发软件的权利。开源软件可以以协作的方式开发。据相关科学家说，开源软件是开放式协作的一个突出的例子。Standish Group 2008 年的一份报告指出，采用开源软件模型每年给消费者节省了约 600 亿美元（约合 480 亿英镑）。

特别是在以下各方面开放源代码具有极高的价值：

安全，负担能力，透明度，永久性，互操作性，灵活性，本地化。

2. Nutch

Nutch is a highly extensible and scalable open source web crawler software project. Nutch is coded entirely in the Java programming language, but data is written in language-independent formats. It has a highly modular architecture, allowing developers to create plug-ins for media-type parsing, data retrieval, querying and clustering. Nutch originated with Doug Cutting, creator of both Lucene and Hadoop, and Mike Cafarella.

Nutch has the following advantages over a simple fetcher:

- Highly scalable and relatively feature rich crawler.
- Features like politeness, which obeys robots.txt rules.
- Robust and scalable — Nutch can run on a cluster of up to 100 machines.
- Quality — crawling can be biased to fetch "important" pages first.

Nutch 是一个高度可扩展、可升级的开源 Web 爬虫软件项目。Nutch 完全以 Java 编程语言编码，但是数据是用语言无关的格式编写的。它具有高度模块化的架构，允许开发人员创建用于媒体类型解析、数据检索、查询和聚类的插件。Nutch 由 Doug Cutting 和 Mike Cafarella 共同开创，其中 Doug Cutting 还是 Lucene 和 Hadoop 的创始人。

Nutch 相对于简单的抓取具有以下优点：

- 高度可扩展性和相对丰富的爬虫功能；
- 规范性，遵守 robots.txt 规则；
- 健壮性和可扩展性——Nutch 可以在多达 100 台机器的集群上运行；
- 质量——爬行偏向于先获取“重要”页面。

3. Distributed File System

Distributed File System (DFS) is a set of client and server services that allow an organization using Microsoft Windows servers to organize many distributed SMB file shares into a distributed file system. DFS provides location transparency (via the namespace component) and redundancy (via the file replication component) to improve data availability in the face of failure or heavy load by allowing shares in multiple different locations to be logically grouped under one folder, or DFS root.

The Distributed File System (DFS) technologies offer wide area network (WAN)-friendly replication as well as simplified, highly-available access to geographically dispersed files. The two technologies in DFS are the following:

DFS Namespaces. Enables you to group shared folders that are located on different servers into one or more logically structured namespaces. Each namespace appears to users as a single shared folder with a series of subfolders.

DFS Replication. DFS Replication is an efficient, multiple-master replication engine that you can use to keep folders synchronized between servers across limited bandwidth network connections.

分布式文件系统（DFS）是客户端和服务器服务的一个组合，允许组织使用 Microsoft Windows 服务器将许多分布式 SMB 文件共享组织到分布式文件系统中。DFS 通过允许在一个文件夹或 DFS 根目录下有逻辑地分组，提供位置透明度（通过命名空间组件）和冗余（通过文件复制组件）来提高面对故障或重负载的数据可用性。

分布式文件系统（DFS）技术提供广域网（WAN）——友好复制以及简化，高可用性访问物理位置分散的文件。

DFS 中的两项技术如下：

DFS 命名空间。DFS 命名空间能够将位于不同服务器上的共享文件夹分组为一个或多个逻辑结构的命名空间。每个命名空间对用户显示为具有一系列子文件夹的单个共享文件夹。

DFS 复制。DFS 复制是一种高效的多主复制引擎，可用于通过有限带宽网络连接在文件夹间保持文件夹同步。

4. High Performance Computing (HPC)

High Performance Computing (HPC) most generally refers to the practice of aggregating computing power in a way that delivers much higher performance than one could get out of a typical desktop computer or workstation in order to solve large problems in science, engineering, or business.

High performance computers of interest to small and medium-sized businesses today are really clusters of computers. Each individual computer in a commonly configured small cluster has between one and four processors, and today's processors typically have from two to four cores. HPC people often refer to the individual computers in a cluster as nodes. A cluster of interest to a small business could have as few as four nodes, or 16 cores. A common cluster size in many businesses is between 16 and 64 nodes, or from 64 to 256.

The point of having a high performance computer is so that the individual nodes can work together to solve a problem larger than any one computer can easily solve. And, just like people, the nodes need to be able to talk to one another in order to work meaningfully together. Of course computers talk to each other over networks, and there are a variety of computer network (or interconnect) options available for business cluster.

高性能计算（HPC）通常指的是将计算能力集合起来的方法，这种方法可以提供比一般台式计算机或工作站更高的性能，以解决科学、工程或商业中的大问题。

目前中小型企业感兴趣的高性能计算机实际上是计算机集群。通常配置的小型集群中的每台计算机都有 1～4 个处理器，而今天的处理器通常有 2～4 个内核。HPC 人员经常将集群中的单个计算机称为节点。一个小企业感兴趣的集群可能只有 4 个节点，即 16 个核。许多企业中的常见集群大小在 16～64 个节点，或从 64～256 个节点。

具有高性能的计算机的要点是各个节点可以一起工作来解决比任何一台计算机能解决问题更大的问题。而且，就像人一样，节点之间需要能够彼此交谈，以便有效工作。当然计算机通过网络相互通信，并且有各种各样的计算机网络（或互联）可供商业集群选择。

5. Storage Area Network (SAN)

A Storage Area Network (SAN) is a dedicated high-speed network(or subnetwork) that interconnects and presents shared pools of storage devices to multiple servers.

A storage-area network is typically assembled using three principle components: cabling, host bus adapters (HBAs) and switches. Each switch and storage system on the SAN must be interconnected and the physical interconnections must support bandwidth levels that can adequately handle peak data activities.

Storage-area networks are managed centrally, and Fibre Channel (FC) SANs have the reputation of being expensive, complex and difficult to manage. The emergence of SCSI has reduced these challenges by encapsulating SCSI commands into IP packets for transmission over an Ethernet connection, rather than an FC connection. Instead of learning, building and managing two networks — an Ethernet local-area network (LAN) for user communication and an FC SAN for storage — an organization can now use its existing knowledge and infrastructure for both LANs and SANs.

存储区域网络（SAN）是一种专用的高速网络（或子网络），它将多个存储设备的共享池互联，并展现给多个服务器。

存储区域网络通常由三个主要组件组成：布线、主机总线适配器（HBA）和交换机。SAN 上的每个交换机和存储系统必须互联，并且物理互联必须支持可以充分处理峰值数据活动的带宽级别。

存储区域网络集中管理，光纤通道（FC）SAN 的价格昂贵、复杂且难以管理。SCSI 的出现减少了这些问题，将 SCSI 命令封装到 IP 数据包中，并通过以太网连接传输，而不是 FC 连接进行传输。企业现在可以将其现有的知识和基础架构用于 LAN 和 SAN，代替掌握、构建和管理两个网络——用于用户通信的以太网局域网（LAN）和用于存储的 FC SAN。

6. Relational DataBase Management System (RDBMS)

A Relational DataBase Management System (RDBMS) is a DataBase Management System (DBMS) that is based on the relational model as invented by E. F. Codd, of IBM's San Jose Research Laboratory. In 2017, many of the databases in widespread use are based on the relational database model.

RDBMS has been a common choice for the storage of information in new databases used for financial records, manufacturing and logistical information, personnel data, and other applications since the 1980s. Relational databases have often replaced legacy hierarchical databases and network databases because they are easier to understand and use. However, relational databases have received unsuccessful challenge attempts by object database management systems in the 1980s and 1990s and also by XML database management systems in the 1990s. Despite such attempts, RDBMS keep most of the market share, which has also grown over the years.

关系数据库管理系统（RDBMS）是基于 IBM San Jose 研究实验室的 E. F. Codd 发明的关系模型的数据库管理系统（DBMS）。在 2017 年，许多广泛使用的数据库都是基于关系数据库模型。

自 20 世纪 80 年代以来，RDBMS 一直是用于财务记录、制造和后勤信息、人事数据和其他应用程序的新数据库中信息存储的常用选择。关系数据库往往取代传统的分层数据库和网络数据库，因为它们更易于理解和使用。然而，关系数据库在 20 世纪 80 年代和 90 年代受到对象数据库管理系统的挑战，在 20 世纪 90 年代也受到 XML 数据库管理系统的挑战。尽管有这样的威胁，RDBMS 还是保住了大部分市场份额，并多年来保持持续增长。

Comprehension

Blank Filling

1. Hadoop has moved far beyond its beginnings in web ________ and is now used in many industries for a huge variety of tasks that all share the common theme of lots of _______, _________ and _________ of data — both structured and unstructured.
2. Using the MapReduce model, Hadoop can take a________ over a dataset, divide it, and run it in parallel over multiple _______. Distributing the computation solves the problem of having data that's too large to fit onto a ________ machine.
3. Apache Hadoop includes a Distributed File System (HDFS), which breaks up ________ data and stores data on the ___________. This makes it possible for data to be processed ___________ using all of the machines in the ________.
4. Hadoop's simplified programming model allows users to quickly write and test software in __________ systems.

Content Questions

1. Why did Google invent a new style of data processing known as MapReduce?
2. Where is Hadoop used?
3. What is the good point of Hadoop?
4. What can Hadoop software stack do?
5. What is the difference between Hadoop and databases?
6. How is Hadoop different from past techniques?
7. Why do we use Apache Hadoop?

Answers

Blank Filling

1. indexing; variety; volume; velocity
2. query; nodes; single
3. input; compute nodes; in parallel; cluster
4. distributed

Content Questions

1. The web was generating more and more information on a daily basis, and it was becoming very difficult to index over one billion pages of content. In order to cope, Google invented a new style of data processing known as MapReduce.

2. It is now widely used across industries, including finance, media and entertainment,

government, healthcare, information services, retail, and other industries with big data requirements but the limitations of the original storage infrastructure remain.

3. The beauty of Hadoop is that it is designed to efficiently process huge amounts of data by connecting many commodity computers together to work in parallel.

4. The Hadoop software stack introduces entirely new economics for storing and processing data at scale. It allows organizations unparalleled flexibility in how they're able to leverage data of all shapes and sizes to uncover insights about their business. Users can now deploy the complete hardware and software stack including the OS and Hadoop software across the entire cluster and manage the full cluster through a single management interface.

5. Unlike databases, Hadoop doesn't insist that you structure your data. Data may be unstructured and schemaless. Users can dump their data into the framework without needing to reformat it. By contrast, relational databases require that data be structured and schemas be defined before storing the data.

6. Hadoop can handle data in a very fluid way.

Hadoop has a simplified programming model.

Hadoop is easy to administer.

Hadoop is agile.

7. It's cost effective. It's fault-tolerant. It's flexible. It's scalable.

参考译文

使用 Hadoop 进行大数据分析

Apache Hadoop 的产生源于对大量数据的处理需求。网络每天都在产生越来越多的信息，而且很难将十多亿页内容进行索引。为了应对这种情况，Google 发明了一种称为 MapReduce 的新型数据处理方式。在 Google 发布了一篇描述 MapReduce 框架的白皮书一年之后，Doug Cutting 和 Mike Cafarella 在白皮书的启发下创建了 Hadoop，将这些概念应用于开源软件框架以支持 Nutch 搜索引擎项目的分发。在最初的情况下，Hadoop 是用简单的一次写入存储基础设施设计的。

Hadoop 已经不限用于最初时的网络索引，现在已经在许多行业的多种任务中使用，这些任务具有共同的特征，即结构化和非结构化数据都具有多样性、数据量大和处理速度快等。它现在广泛应用于包括金融、娱乐媒体、政府、医疗保健、信息服务、零售和其他具有大数据需求的行业，但原有存储基础设施的局限性仍然存在。

Hadoop 正日益成为大规模、数据密集型部署的框架。Hadoop 用于处理数据量从太字节（TB）到拍字节（PB）级及以上的大量数据。因为数据量庞大，它不太适合于单个计算机的硬盘驱动器和内存。Hadoop 的优点在于它通过将许多商用计算机连接在一起并行工作来高效地处理大量数据。使用 MapReduce 模型，Hadoop 可以对数据集进行查询，将其划分并在多个节点上并行运行。分布计算解决了数据太大而不能适应单个机器的问题。

Hadoop 软件

Hadoop 软件堆栈为成比例的数据存储和处理提供了全新的经济性。它允许企业在如何利用各种数据方面具有无与伦比的灵活性以发掘有关其业务的见解。用户可以在整个集群中部署完整的硬件和软件堆栈，包括 OS 和 Hadoop 软件，并通过一个管理界面管理完整的集群。

Apache Hadoop 包括了分布式文件系统（HDFS），它将输入的数据分解并将其存储在不同的计算节点上，这种处理方式可以使用集群中的所有机器并行处理数据。Apache Hadoop 是用 Java 编写的、可以在不同的操作系统上运行的分布式文件系统。

Hadoop 从一开始就被设计为可适应多个文件系统，并且有一些已经实现。HDFS 和 S3 文件系统可能是最广泛使用的，但还有许多其他可用的文件系统如 MapR 文件系统。

Hadoop 与过去的技术有何不同

Hadoop 可以以非常流畅的方式处理数据。Hadoop 不仅仅是一个更快、更便宜的数据库和分析工具。与（传统）数据库不同，Hadoop 并不强制构建数据结构。数据可能是非结构化的和无模式的。用户可以将其数据转储到框架中而无须重新格式化。相比之下，关系数据库需要在存储数据之前对数据进行结构化和模式定义。

Hadoop 具有简化的编程模型。Hadoop 的简化编程模型允许用户在分布式系统中快速编写和测试软件。在分布式环境中执行大量数据计算已经实现，但为分布式系统编写软件是非常困难的。通过交换一些编程灵活性，Hadoop 使编写分布式程序变得更加容易。

由于 Hadoop 几乎可以接受任何类型的数据，它存储的信息要比通常在传统数据库中整齐的行和列中找到的格式多得多。例如，包含 JSON、Avro 和 ORC 的存储格式编写的机器生成的数据和日志数据。

Hadoop 中的大部分数据准备工作是使用脚本语言（如 Hive、Pig 或 Python）编写代码。

Hadoop 易于管理。备选的高性能计算（HPC）系统允许程序在大型计算机系统上运行，但它们通常需要严格的程序配置，并且通常需要将数据存储在单独的存储区域网络（SAN）系统上。HPC 集群上的调度程序需要仔细管理，并且由于程序执行对节点故障敏感，因此对 Hadoop 集群的管理要容易得多。

Hadoop 能够隐性地处理诸如节点故障之类的作业控制问题。如果节点出现故障，Hadoop 将确保在其他节点上运行计算，并且从其他节点恢复存储在该节点上的数据。

Hadoop 是敏捷的。关系数据库擅长用预定义的和严格的数据模型存储和处理数据集。对于非结构化数据，关系数据库缺乏其所需的敏捷性和可扩展性。Apache Hadoop 通过将大量的结构化和非结构化数据一起处理和分析来降低成本，并且不用提前定义结构进行处理数据。

为什么要使用 Apache Hadoop

成本低。Apache Hadoop 通过比其他平台更经济实惠的每太字节（TB）存储数据来控制成本。Hadoop 的每太字节（TB）只需数百美元（并非数千到数万美元）来计算和存储。

容错性。容错是使用 Hadoop 最重要的优点之一。虽然单个节点在大型集群上运行作业时有很高的故障率，数据也可被跨集群复制，以便在磁盘、节点或机架故障时轻松恢复。

灵活性。数据存储在 Apache Hadoop 中的灵活方式是其最大的优势之一，使企业能够从先前被认为过于昂贵而不能在传统数据库中进行存储和处理的数据中生成价值。使用 Hadoop 可以使用所有类型的数据——结构化和非结构化，从中提取更有意义的商业洞察力。

可扩展性。Hadoop 是一个高度可扩展的存储平台，因为它可以在数以百计的并行运行的廉价服务器的集群中存储和分配非常大的数据集。传统的关系数据库管理系统（RDBMS）的问题是它们无法扩展到处理大量数据。

Chapter 4

Impacts of Big Data

Text A

Big data is both a marketing and a technical term referring to a valuable enterprise asset — information. Big data represents a trend in technology that is leading the way to a new approach in understanding the world and making business decisions. These decisions are made based on very large amounts of structured, unstructured and complex data (e.g., tweets, videos, commercial transactions) which have become difficult to process using basic database and warehouse management tools. Managing and processing the ever-increasing data set requires running specialized software on multiple servers. For some enterprises, big data is counted in hundreds of gigabytes; for others, it is in terabytes or even petabytes, with a frequent and rapid rate of growth and change (in some cases, almost in real time). In essence, big data refers to data sets that are too large or too fast-changing to be analyzed using traditional relational or multidimensional database techniques or commonly used software tools to capture, manage and process the data at a reasonable elapsed time.

Data are collected to be analyzed to find patterns and correlations that may not be initially apparent, but may be useful in making business decisions. This process is called big data analytics. These data are often personal data that are useful from a marketing perspective in understanding the likes and dislikes of potential

New Words and Expressions

ever-increasing data
不断增长的数据

multidimensional *adj.* /ˌmʌltidaɪˈmenʃənəl/
多面的，多维的

elapse /ɪˈlæps/ *v.*
消逝；时间过去

apparent /əˈpærənt/ *adj.*
显然的；貌似的，表面的

buyers and in analyzing and predicting their buying behavior. Personal data can be categorized as:

- Volunteered data — Created and explicitly shared by individuals (e.g., social network profiles)
- Observed data — Captured by recording the actions of individuals (e.g., location data when using cell phones)
- Inferred data — Data about individuals based on analysis of volunteered or observed information (e.g., credit scores)

The primary objective of analyzing big data is to support enterprises in making better business decisions. Data scientists and other users analyze large amounts of transaction data as well as other data sources that may be ignored by traditional business intelligence software[1], such as web server logs, social media activity reports, cell phone records and data obtained via sensors. Data analytics can enable a targeted marketing approach that gives the enterprise a better understanding of its customers — an understanding that will influence internal processes and, ultimately, increase profit, which provides the competitive edge most enterprises are seeking.

Impact of Big Data on the Enterprise

Big data can impact current and future process models in many ways. Beyond a business impact, the aggregation of data can affect governance and management over planning, utilization, assurance and privacy:

- **Governance** — What data should be included and how should governance of big data be defined and delivered?
- **Planning** — Planning involves the process of collecting and organizing outcomes to:

– Justify process adjustments or improvements which until recently could be identified using specialized research techniques such as predictive modeling[2].

– Design a trading program predicated on certain conditions that trigger events.

– Encourage target purchase patterns while a buyer is researching products and services.

– Use location-based information in combination with other collected data to guide customer loyalty, route traffic, identify new product demands, etc.

New Words and Expressions

profile/ˈprəʊfaɪl/ *n.*
侧面；外形，轮廓；人物简介

competitive edge
竞争优势

predictive modeling
预测建模

aggregation/ˌægrɪˈgeɪʃən/ *n.*
聚集；集成；集结

– Manage just-in-time (JIT) inventory[3] based on seasonal or demand changes. For example, a manufacturing enterprise may adjust production levels for a particular item after the part number is not ordered for two consecutive days.

– Manage operations of logistics and transportation firms based on real-time performance[4].

– Manage unplanned IT infrastructure and policy changes that disrupt the direction of IT support.

- **Utilization** — Use of big data can vary from one enterprise to another depending on the enterprise's culture and maturity. A small enterprise may be slower to adopt big data because it may not have the necessary infrastructure to support the new processes involved. Companies such as IBM, Hewlett-Packard Company (HP) and Amazon.com, on the other hand, have changed direction over the last few years from selling products to providing services and using information to guide business decisions. Companies that have embraced big data have made the necessary investments to become information mavens capable of identifying new product and service demands using data mining — information that they then turn into a competitive advantage by being the first to market.

Infrastructures built to support big data are also cross-marketed to support cloud computing services, in a way making customers business partners (causing the rise of phrases such as "frenemies" and "coopetition"). In other words, big data customers may be competitors in one geometric plane and cooperative partners in another, as with Netflix using the Amazon.com cloud infrastructure to support its media streaming[5].

- **Assurance** — Experience leads enterprises to develop better assurance practices. Once leadership develops a strategy that leverages big data, the enterprise can focus on defining an assurance framework to control and protect big data. The main concern for the assurance organization is data quality, addressed by topics such as normalization, harmonization and rationalization.

New Words and Expression

just-in-time inventory
适时存货

real-time performance
实时性能

consecutive/kənˈsekjətɪv/ *adj*
连续的，连贯的

maturity/məˈtʃʊərəti/ *n.*
成熟；完备；（票据等的）到

maven/ˈmeɪvən/ *n.*
<美口>专家，内行

frenemy/ˈfrenəmi/
=friend+enemy 友敌；指伪装成友的敌人或者互相竞争的同伴用来指代个人及群体组织之间人际关系、地缘政治关系以及商关系

coopetition/kəʊˌɒpərˈeɪʃən/
=cooperate+competition
合作竞争

media streaming
流媒体

normalization /ˈnɔːməlaɪˈzeɪʃn/
正常化；标准化

harmonization
/ˈhɑːmənaɪˈzeɪʃn/ *n.*
和谐，协调，相称

rationalization
/ˈræʃənəlɪzeʃən/ *n.*
合理化，合于经济原则

- **Privacy** — Privacy protection has always been handled differently by geographic regions, governments and enterprises. Laws protect the privacy of individuals and any information collected about them, even if people share confidential information inappropriately, for example, posting nonpublic or private information (e.g., pictures of credit cards, birthdays, phone numbers, personal preferences) in social media outlets. Regardless of the authenticity of information collected from social media, its collection requires protection from nefarious users as well as over-controlling governments.

New Words and Expressions

personal preference
个人爱好
outlet/ˈaʊtˌlet/ *n.*
出口，出路；批发商店
nefarious/nəˈfeəriəs/ *adj.*
极坏的，恶毒的

Note:

The text is adapted from the website:

http://www.isaca.org/Knowledge-Center/Research/ResearchDeliverables/Pages/Big-Data-Impacts-and-Benefits.aspx.

Terms

1. Business intelligence software

Business intelligence software is a type of application software designed to retrieve, analyze, transform and report data for business intelligence. The applications generally read data that have been previously stored, often, though not necessarily, in a data warehouse or data mart.

In the years after 2000, business intelligence software producers became interested in producing universally applicable BI systems which don't require expensive installation, and could be considered by smaller and midmarket businesses which could not afford on premise maintenance. These aspirations emerged in parallel with the cloud hosting trend, which is most vendors came to develop independent systems with unrestricted access to information.

商务智能（Business Intelligence，BI）软件是一种用于检索、分析、转换和报告商业智能数据的应用软件。该应用程序通常（但不一定）在数据仓库或数据中心中读取预先存储的数据。

在 2000 年以后的几年里，商务智能软件生产商开始对生产普遍适用的 BI 系统感兴趣，这些 BI 系统不需要在安装上花费太多，因此无法负担内部维护的中小型企业可以考虑使用。这些愿景与云托管趋势并行出现，这将使大多数服务供应商开发一种可以无限制地访问信息的独立系统。

2. Predictive modeling

Predictive modeling uses statistics to predict outcomes. Most often the event one wants to predict is in the future, but predictive modeling can be applied to any type of unknown event, regardless of when it occurred. For example, predictive models are often used to detect crimes and identify suspects, after the crime has taken place. The applications of predictive modeling are shown as Figure 4-1.

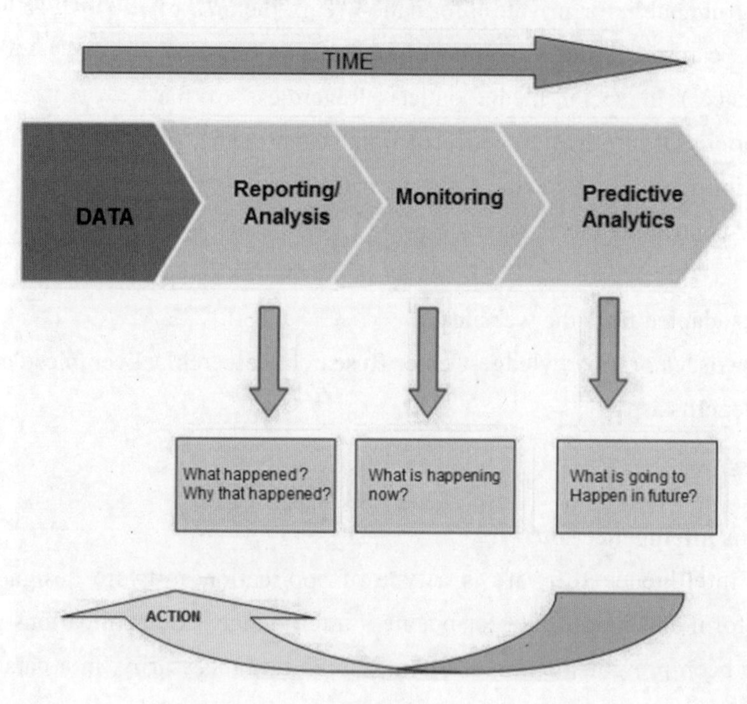

Figure 4-1

Archaeology

Predictive modeling in archaeology gets its foundations from Gordon Willey's mid-fifties work in the Virú Valley of Peru. Complete, intensive surveys were performed then co-variability between cultural remains and natural features such as slope, and vegetation were determined.

Customer relationship management

Predictive modeling is used extensively in analytical customer relationship management and data mining to produce customer-level models that describe the likelihood that a customer will take a particular action. The actions are usually sales, marketing and customer retention related. For example, a large consumer organization such as a mobile telecommunications operator will have a set of predictive models for product cross-sell, product deep-sell and churn.

Health care

In 2009 Parkland Health & Hospital System began analyzing electronic medical records in order to use predictive modeling to help identify patients at high risk of readmission. Initially the hospital focused on patients with congestive heart failure, but the program has expanded to include patients with diabetes, acute myocardial infarction, and pneumonia.

预测建模使用统计来预测结果。大多数情况下，人们想要预测的事都发生在未来，但预测模型可以应用于任何类型的未知事件，无论这些未知事件何时发生。例如，犯罪发生后，预测模型通常用于检测犯罪和识别嫌疑犯。预测模型的应用如图 4-1 所示。

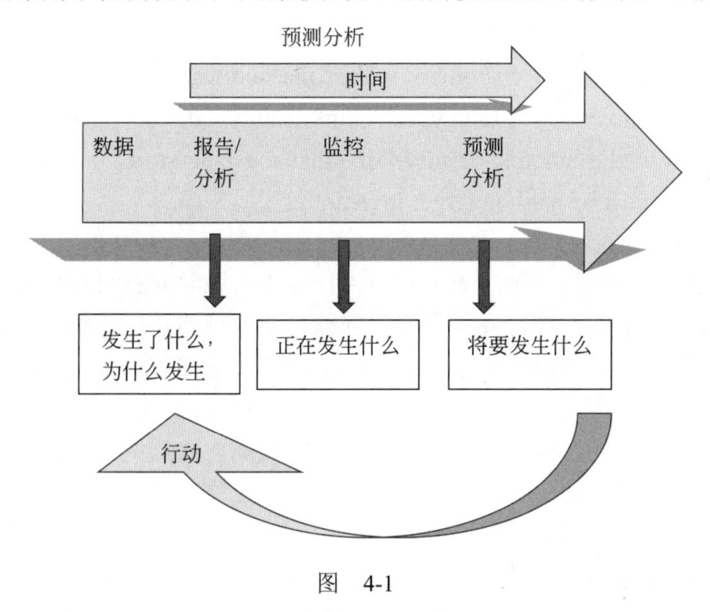

图 4-1

考古学

19 世纪 50 年代中叶，戈登 • 威利（Gordon Willey）在秘鲁维鲁山谷的工作为考古学中的预测建模奠定了基础。他进行了全面深入的调查，确定了文化遗产与斜坡、植被等自然特征的共变性。

客户关系管理

预测建模被广泛用于分析客户关系管理和数据挖掘，以生成描述客户可能采取某种特定行动的客户级模型。这些行为通常与销售、营销和客户保留相关（Customer Retention，企业为防止客户流失和提高客户忠诚度所建立的一整套策略和方法。保留一个老客户的成本是获取一个新客户成本的 1/5，几乎所有的销售人员都会知道向一个现有客户销售产品要比不断寻求新客户容易得多）。例如，移动电信运营商等大型消费者群体将拥有一套客户关系管理的整合产品，包括交叉销售、产品深度销售和流失的预测模型。

保健

2009 年，Parkland Health＆Hospital System 开始分析电子医疗记录，以便使用预测模型来帮助识别那些再入院风险高的患者。最初，医院专注于充血性心力衰竭患者，但该计划已扩大到包括糖尿病、急性心肌梗死和肺炎患者。

3. Just-In-Time (JIT) inventory

JIT, or just in time, inventory is an inventory management strategy that is aimed at monitoring the inventory process in such a manner as to minimize the costs associated with inventory control and maintenance. To a great degree, a just-in-time inventory process relies on the efficient monitoring of the usage of materials in the production of goods and ordering replacement goods that arrive shortly before they are needed. This simple strategy helps to prevent incurring the costs associated with carrying large inventories of raw materials at any given point in time.

Another application of a just in time inventory focuses not on raw materials but on finished goods. The idea is to develop a solid understanding of what is needed to produce goods and schedule them for shipment to customers within the shortest time frame possible. This dual application of a JIT inventory strategy can significantly cut the operational expenses of a business in regards to the amount of inventory that must be stored at any one time and the amount of taxes that must be paid on larger inventories.

JIT（即时库存）是一种库存管理策略，指在监控库存的过程中，以将库存控制和维护的相关成本最小化为目标的一种方式。在很大程度上，即时库存过程依赖于对货物生产中的材料的使用进行有效的监控，并且在库存红线之前就订购替换货物。这种简单的策略有助于防止在任何给定时间点储存大量原材料，造成相关的储存成本损失。

即时库存的另一个应用不是聚焦于原材料，而是在成品上。该计划是对生产商品所需要的原材料有一个深入的认识，并且可以在最短的时间内将这些原材料运送给客户。无论何时，JIT 库存策略的双重应用都可以显著降低企业必须存储的库存量的运营费用，以及在较大库存中必须支付的税额。

4. Real-time performance

In computer science, real-time computing (RTC), or reactive computing describes hardware and software systems subject to a "real-time constraint", for example from event to system response. Real-time programs must guarantee response within specified time constraints, often referred to as "deadlines". The correctness of these types of systems depends on their temporal aspects as well as their functional aspects. Real-time responses are often understood to be in the order of milliseconds, and sometimes microseconds. A system not specified as operating in real time cannot usually guarantee a response within any timeframe, although actual or expected response times may be given.

Real-time software may use one or more of the following: synchronous programming languages, real-time operating systems, and real-time networks, each of which provide essential frameworks on which to build a real-time software application. Systems used for many mission critical applications must be real-time, such as for control of fly-by-wire aircraft, or anti-lock brakes on a vehicle, which must produce maximum deceleration but intermittently stop braking to prevent skidding. Real-time processing fails if not completed within a specified deadline relative to an event; deadlines must always be met, regardless of system load.

在计算机科学中，实时计算（RTC）或反应计算描述了受到"实时约束"的硬件和软件系统，例如从事件到系统响应的过程。实时程序必须保证在规定时间内的响应，这通常被称为"最后期限"。这类系统的正确性取决于它们的时间是否实时以及它们的功能。实时响应通常被理解为毫秒级，有时是微秒级。虽然可以给出实际或预期的响应时间，但未被确定为实时操作的系统通常不能保证在任何时间范围内的响应。

实时软件可以使用以下的一个或多个功能：同步编程语言、实时操作系统和实时网络，每个网络都提供构建实时软件应用程序的基本框架。用于多任务关键应用的系统必须是实时的，例如，用于控制飞行中的飞机或车辆上的必须产生最大减速但间歇地停止制动以防止打滑的防抱死制动器。如果在相对于事件的规定期限内未完成，则实时处理失败；无论系统负载如何，都必须始终满足最后期限。

5. Media streaming

Streaming media is multimedia that is constantly received by and presented to an end-user while being delivered by a provider. The verb "to stream" refers to the process of delivering or obtaining media in this manner; the term refers to the delivery method of the medium, rather than the medium itself, and is an alternative to file downloading, a process that the end-user obtains the entire file for the content before watching or listening to it.

A client end-user can use their media player to begin to play the data file (such as a digital file of a movie or song) before the entire file has been transmitted. Distinguishing delivery method from the media distributed applies specifically to telecommunications networks, as most of the delivery systems are either inherently streaming (e.g. radio, television) or inherently non-streaming (e.g. books, video cassettes, audio CDs). For example, in the 1930s, elevator music was among the earliest popularly available streaming media; nowadays Internet television is a common form of streamed media. The term "streaming media" can apply to media other than video and audio such as live closed captioning, ticker tape, and real-time text, which are all considered "streaming text".

流媒体是由供应商交付、不断接收内容并呈现给最终用户的多媒体。动词“流”是指以这种方式传递或获取媒体的过程；该术语是指介质的传递方法，而不是介质本身，是文件下载的替代方案，是终端用户在观看或收听内容之前获取文件内容的过程。

在整个文件传输完毕之前，客户终端用户就可以使用其媒体播放器播放数据文件（如电影或歌曲的数字文件）。该分发方法特别适用于电信网络，因为大多数传送系统是固有的流式传输（如，无线电、电视）或固有的非流传输（如，书籍、录像带、音频 CD）。例如，在 20 世纪 30 年代，电梯音乐是最早流行的流媒体之一；现在互联网电视是流媒体的常见形式。术语“流媒体”也可以应用于视频和音频以外的媒体，例如现场隐藏式字幕、自动代码磁带和实时文本，这些都被视为“流文本”。

Comprehension

Blank Filling

1. Big data is both a _________ and a ________ term referring to a valuable enterprise asset — information. Big data represents a trend in technology that is leading the way to a new approach in _________________ and _____________.
2. Managing and processing the ever-increasing data set requires running specialized _________ on multiple _________.
3. In essence, big data refers to data sets that are too ________ or too _________ to be analyzed using traditional relational or multidimensional database techniques or commonly used software tools to ________, ________ and ________ the data at a

reasonable elapsed time.

4. Data are collected to be analyzed to find ________ and __________ that may not be initially apparent, but may be useful in making business decisions.
5. Big data can impact current and future process models in many ways. Beyond a business impact, the aggregation of data can affect __________ and ___________ over planning, ________, ________ and ________.

Content Questions

1. On what are business decisions made?
2. From a marketing perspective, what are data used for?
3. What is volunteered data?
4. What is observed data?
5. What is inferred data?

Answers

Blank Filling

1. marketing; technical; understanding the world; making business decisions
2. software; servers
3. large; fast-changing; capture; manage; process
4. patterns; correlations
5. governance; management; utilization; assurance; privacy

Content Questions

1. Business decisions are made based on very large amounts of structured, unstructured and complex data (e.g., tweets, videos, commercial transactions) which have become difficult to process using basic database and warehouse management tools. The analytical findings can lead to more effective marketing, new revenue opportunities, better customer service, improved operational efficiency, competitive advantages over rival organizations and other business benefits.
2. Data are collected to be analyzed to find patterns and correlations that may not be initially apparent, but may be useful in making business decisions. This process is called big data analytics. These data are often personal data that are useful from a marketing perspective in understanding the likes and dislikes of potential buyers and in analyzing and predicting their buying behavior.
3. Volunteered data is the data created and explicitly shared by individuals (e.g., social network profiles).
4. Volunteered data is the data captured by recording the actions of individuals (e.g., location data when using cell phones).
5. Inferred data is the data about individuals based on analysis of volunteered or observed information (e.g., credit scores).

参考译文

大数据既是营销术语又是技术术语，指的是有价值的企业资产——信息。大数据代表了技术趋势，正在引领使用新方法来了解世界并做出业务决策。这些决策制定是基于使用非常大量的结构化、非结构化的复杂数据（例如推文、视频、商业交易），这些复杂的数据是传统数据库和数据仓库管理工具都难以应付的。管理和处理不断增长的数据集需要在多台服务器上运行专门的软件。对于一些企业来说，大数据计算量以吉字节（GB）为单位；对于其他人来说，它可以是以太字节（TB）甚至拍字节（PB）为单位，且频繁和快速地增长和变化（在某些情况下，几乎是实时的）。实质上，大数据是指在一个合理的时间内使用传统的关系或多维数据库技术或常用的软件工具来捕获、管理和处理数据量过大或变化过快的数据集，从而进行分析。

收集数据进行分析以找出最初可能不是显而易见的模式和相关性，这有助于实体做出业务决策。这个过程称为大数据分析。这些数据通常是从营销角度来理解潜在买家的喜好和非喜好以及分析和预测其购买行为的个人数据。个人数据可分为：

- 自愿提供的数据——由个人创建和明确共享（例如，社交网络配置文件）；
- 观察数据——通过记录个人的行为获得（例如，使用手机时的位置数据）；
- 推测数据——关于个人的数据基于对自愿或观察信息（例如，信用评分）的分析。

大数据分析的主要目标是支持企业做出更好的业务决策。数据科学家和其他用户分析大量交易数据以及传统商业智能软件可能忽略的其他数据源，如 Web 服务器日志、社交媒体活动报告、手机记录和通过传感器获得的数据。数据分析可以实现有针对性的营销手段，使企业更好地了解其客户，这一理解将影响内部流程，并最终增加利润，从而提供大多数企业正在寻求的竞争优势。

大数据对企业的影响

大数据可以在许多方面影响企业的现在和未来。除了业务影响之外，数据的聚合可能会影响到规划、利用、担保和隐私。

治理：应包括哪些数据，对大数据治理如何定义以及如何交付？

规划：运营数据的收集和组织的过程。

- 通过诸如预测建模等专门的研究技术来保证过程调整和改进的有效性。
- 根据触发事件的某些条件设计交易程序。
- 在买家还在研究产品和服务时，寻找能够鼓励消费的模式。
- 使用基于位置的信息与其他收集的数据相结合，来引导客户忠诚度、路线流量、新产品需求等。
- 根据季节或需求变化管理即时（JIT）库存。例如，当某型号的零件在连续两天没有订购的情况下，制造企业会以此来调整生产。
- 根据实时业绩管理物流运输公司的运营。
- 对影响信息技术基础架构和策略的变化进行管理。

使用：根据企业的文化和成熟度，大型数据的使用可能因企业而异。小型企业采用大数据的速度会比较慢，因为它可能没有必要的基础设施来支持所涉及的新流程。另外，诸如 IBM、HP 和亚马逊等公司已经在过去几年把方向从销售产品转变为提供服务和使用信息来指导业务决策。拥有大量数据的公司已经进行了必要的投资，能够使用数据挖掘信息来识别新产品和服务需求，然后通过首次推向市场将其变成竞争优势。

支持云计算服务的大数据基础设施是服务交叉市场的，(导致诸如“亦敌亦友”和“合作竞争”等词语的兴起)。换句话说，大数据客户可能一面是竞争对手，另一面是合作伙伴，Netflix 则使用亚马逊云基础设施来支持其媒体流。

保障：企业依据经验来制定更好的措施来保障实践。一旦领导层制定了利用大数据的战略，企业就可以专注于保障框架来控制和保护大数据。提供保障的机构主要关注的是数据质量，规范化、协调化和合理化的来解决问题。

隐私：对于不同的地域、政府和企业，他们对待隐私保护的处理方式一直不同。法律保护个人的隐私以及从个人收集的任何信息，包括人们无意识地分享出的机密信息，例如，在社交媒体上发布非公开或私人信息(信用卡、生日、电话号码、个人喜好的图片等)。无论从社交媒体收集到的信息是否真实，这些收集的信息都需要保护其不受恶意用户的影响。

Text B

If there are pros to big data, there are cons too. The negative impact of big data is subtly hidden in the trail of digital traces we unknowingly leave.

The hype around big data is undeniable, especially since organizations have felt the vast benefits of it. Today, every industry uses big data as a powerful tool to serve various different purposes. As industries delve deeper into the landscape of big data, it is crucial for them to grasp the possibility that if deployed carelessly, the negative impact of big data is a real possibility.

Negative impact of big data on privacy and security

Data is easily accessible now days. A single click and a pool of diverse information is presented to us. However, has the thought ever occurred to you that more than half of this data is the amassment of human personal information? A simple action, like logging into social media sites or inputting private details in banks and hospitals for compliance purposes, can leave a digital trail. Just imagine, this happens on a global level and at a continuous pace.

However, what we don't realize is that this information can be more valuable than we think. Let's get ourselves to accept that the possibility for negative impact of big data is always around the corner. By agreeing to reveal personal data, our right to privacy and security is compromised and we become susceptible to data breaches. Incidents of people falling prey to scams that offer a huge amount of money or a stalker who gets access to an address because

New Words and Expressions

negative/ˈnegətɪv/ *adj.*
否定的，有害的，不良的，消极的

trail/treɪl/ *n.*
痕迹，踪迹，路径

hype/haɪp/ *n.*
大肆宣传，炒作

delve/delv/ *v.*
钻研；探究

accessible/əkˈsesəb(ə)l/ *adj.*
可进入的；易得到的，可使用的；易懂的；平易近人的，随和的；易受影响的

amassment/əˈmæsmənt/ *n.*
积蓄；聚积

compliance/kəmˈplaɪəns/ *n.*
服从，遵守

compromise/ˈkɒmprəmaɪz/ *n./ v.*
折中，妥协

susceptible/səˈseptəb(ə)l/ *adj.*
易得病的，易受影响的

of a profile being public, are incidents that are not unheard of. The negative impact of big data has reached huge industries, like in the case of retail and banking; these industrial sectors are known to scrutinize customer behavior and user activity to increase their market reach and acquire financial profits.

Negative impact of big data on data usage

Most people are unaware of the ways in which data can be misused, until they become victims of data manipulation or even worse, a theft or a fraud. While we may casually enter a store or sit down to order from our favorite site, we have no idea that the store or company already has a fair idea about our preferences. They hold insights on our buying capacity and our market knowledge. No wonder the products we seem to fancy most, but are out of our budget, are always on sale!

But what about more critical fields like healthcare and insurance companies? They have direct access to confidential information. Let's consider hackers and data thefts. It seems like a far-off possibility until it actually occurs. Data, when it reaches the wrong hands, can hack down an entire organization or government. Thus, it is crucial that organizations and governments take all possible steps and measures to curb the negative impact of data as much as possible.

Negative impact of big data on society

Those who are tech-savvy and privileged have higher chances of staying safe. But what about individuals who are under-resourced or lack the technical know-how? There's always a risk for them to fail at concepts like risk analysis and data scoring. For instance, banks and insurance companies track the payment history of customers and accordingly reduce or increase their credit limit. This way, the rich and privileged experience more luxury, while the middle class ends up becoming victims of big data negativity.

Talking about risk analysis, the best and most recent example would be the use of big data for regional and national security purposes. While governments are in the pursuit of detecting terrorists, they might end up discriminating against people of a certain race or religion. Consequently, innocent lives may end up behind bars. One simply cannot ignore the adverse effect of such stratification and the negative impact of big data on innocent people.

New Words and Expressions

scrutinize/ˈskruːtənaɪz/ *v.*
仔细查看

manipulation/məˌnɪpjuˈleɪʃn/ *n.*
控制，操作，处理

confidential/ˌkɒnfɪˈdenʃl/ *adj.*
机密的，保密的

curb/kɜːrb/ *v.*
控制，抑制

tech-savvy/ˌtek ˈsævi/ *adj.*
精通技术的

privileged/ˈprɪvəlɪdʒd/ *adj.*
享有特权的

know-how/ˈnəʊ haʊ/ *n.*
〈非正式〉专门知识，实际经验

adverse/ˈædvɜːs/ *adj.*
不利的，有害的；相反的

Anything that is unmonitored leaves an opportunity for exploitation. The same is the case with big data. While big data is not bad in itself, it can have undesirable effects if the people involved in its use have malicious intentions. It is time that individuals and organizations become aware of the value personal data and information holds and adopt a more transparent approach.

New Words and Expressions

malicious/mə'lɪʃəs/ *adj.* 恶意的

Note:

The text is adapted from the website:

https://resources.experfy.com/bigdata-cloud/how-big-data-can-cause-negativity/

参考译文

大数据有优点，同时它也有缺点。我们在不知不觉中留下数字痕迹，大数据的负面影响就隐藏在其中。

不可否认，对于大数据存在一定的炒作，特别是很多组织已经感受到了大数据带来的巨大好处。今天，每个行业都将大数据作为一个强大的工具，来提供不同的服务。随着各行业对大数据的深入研究，大家要认识到它的另一种可能性：如果处理不当，大数据的负面影响是真实存在的。

大数据对隐私和安全的不利影响

现在的数据很容易获得，只需点击一下鼠标，就会有大量不同的信息呈现在我们面前。然而，你有没有想过这些数据中有一半以上会涉及个人信息？一个简单的动作，如登录社交媒体网站或在银行和医院输入个人的详细信息，都可能留下数字痕迹。试想一下，这样的事情在全世界不断地发生着。

然而，我们没有意识到的是，这些信息可能比我们想象的更有价值。我们要认识到大数据的负面影响是客观存在的。在填写个人数据时，我们的隐私和安全便受到了损害，我们的数据就存在泄露的风险。有的人成了巨额诈骗的牺牲品，有的因个人资料被公开而被跟踪，这些事件并不罕见。大数据的负面影响已经波及某些行业，比如零售业和银行业，这些行业为扩大市场和获取利润，都需要仔细审核客户行为和用户活动。

大数据对数据使用的消极影响

大多数人直到成为受害者才意识到他们的数据可能被滥用，更糟糕的是，有的人甚至被盗窃或欺诈。当我们进入一家商店或在我们最喜欢的网站上购物时，我们并不知道商店或公司已经对我们的喜好、购买能力和市场知识已经有了深入的了解。这就是为什么超出预算自己却很喜欢的商品总是在打折!

但是，像医疗和保险这样更重要的领域呢？他们可以直接接触到机密信息，让我们看一下这些领域遭遇黑客和数据盗窃会是什么后果。当这些没发生时，人们会觉得很遥远。数据一旦落入坏人之手，就能黑掉整个组织或政府。因此，组织和政府应该采取一切可能的措施尽可能地遏制数据的负面影响。

大数据对社会的消极影响

对于那些精通技术和地位高的人来说，他们的数据会更安全。但那些资源不足或缺乏技术的人呢？他们并不能理解风险分析和数据评分这类概念。例如，银行和保险公司会跟踪客户的付款历史，来相应地减少或增加其信用额度，这样一来，富人会有更多的额度用于奢侈消费，而中产阶级则会受到大数据的消极影响。

谈到风险分析，最好也是最近的例子就是将大数据用于区域和国家安全。当政府在追踪和侦查恐怖分子时，他们可能会歧视某个种族或宗教的人。因此，无辜的人可能会被关进监狱，人们不能忽视大数据进行这种分类对无辜者产生的负面影响。

任何不受监控的东西都会留下被利用的机会，大数据的情况也是如此。虽然大数据本身并不坏，但如果参与使用大数据的人有恶意，它就会产生不良影响，现在是到了该重视个人数据和信息价值并采取行动的时候了。

Chapter 5

Business Benefits of Big Data

Text A

Businesses are using the power of insights provided by big data to instantaneously establish who did what, when and where. The biggest value created by these timely, meaningful insights from large data sets is often the effective enterprise decision-making that the insights enable.

Big data opportunities are significant, as are the challenges. Enterprises that master the emerging discipline of big data management can reap significant rewards and differentiate themselves from their competitors. Indeed, research conducted by Erik Brynjolfsson, an economist at the Sloan School of Management at the Massachusetts Institute of Technology (USA), shows that companies that use "data-directed decision making[1]" enjoy a five to six percent boost in productivity. Proper use of big data goes beyond collecting and analyzing large quantities of data; it also requires understanding how and when to use the data in making crucial decisions.

Competitive advantage can be greatly improved by leveraging the right data. According a research report by McKinsey, the potential value from data in the US health care sector could be more than US $300 billion in value every year, two-thirds of which would be in the form of reducing national health care expenditures by approximately eight percent.

Financial benefits can be realized when data management

New Words and Expressions

instantaneous/ˌɪnstənˈteɪniəs/ *adj*.
瞬间的；即刻的；猝发的

boost/buːst/ *n*./*v*.
促进，提高；增加；吹捧

crucial/ˈkruːʃəl/ *adj*.
关键性的，极重要的；决定性的

processes are aligned with the enterprise's strategy, which may require top management involvement to set direction and oversee major decisions.

Big data analytics can positively impact:

- Product development
- Market development
- Operational efficiency
- Customer experience and loyalty
- Market demand predictions

Big Data is a buzzword amongst businessmen nowadays. Regardless of industry or company size, it manages to squeeze into every nook and cranny. There are at least three ways that Big Data has been impacting companies that everyone should understand moving forward.

It has revolutionized old-school industries.

"Big Data has had a tremendous impact on businesses from customer relations to supply chain operations and will continue to do so" says Edwin Miller, CEO of 9Lenses. While many may not be aware of it, Big Data has indeed made a significant impact on many traditional businesses.

Capital One is a perfect example. In the 1990s, the credit card industry utilized a uniform-pricing model charging every customer the same price, with the exception of Capital One. The company used a statistical model[2] based on public credit and demographic data to provide customers with "custom-tailored" products. The innovation was one of their cornerstone developments in earning 32% CAGR in net revenue (after provisions) from 1994 to 2003. Consequently, many banks have shifted focus towards Big Data analytics, but the pioneers seem to have maintained their edge. Their annual net revenue has increased by 17% compared with top banks in the US such as Citigroup at 11%, Bank of America at 11% and JP Morgan at 6% from 2009 to 2014.

Rolls Royce's success in applying Big Data analytics has influenced the aircraft engine-manufacturing sector. The company consistently monitors approximately 3,700 engines, each of which has hundreds of censors installed, to predict when and where breakdowns may occur. Roll-Royce has transformed from selling only engines to selling packages of both engines and monitoring

New Words and Expressions

align/əˈlaɪn/ *v.*
使成一线，使结盟；排列

nook/nʊk/ *n.*
角落；隐蔽处；每个角落；到处

cranny/ˈkræni/ *n.*
裂缝，裂隙

revolutionize/ˌrevəˈluːʃənaɪz/ *v.*
发动革命；彻底改革；使革命化

tremendous /trɪˈmendəs/ *adj.*
极大的，巨大的；可怕的；极好的

demographic/ˌdeməˈgræfik/ *adj.*
人口统计学的；人口统计的

cornerstone/ˈkɔːnəstəʊn/ *n.*
奠基石；基石；最重要的部分

services and then charging customers based on usage, repairs and replacements. The service currently accounts for more than 70% of their annual revenue in their aircraft engine division.

Another perfect example is the world's largest retailer, Walmart(Figure 5-1). Walmart is a well-known user of Big Data analytics today, but in the 1990s, it reformed the retail industry by recording every product as data through a system called Retail Link. The system provided a way for suppliers to manage their own products by allowing them to monitor their data, including sales and inventory volume, in-stock percentage, gross margin and inventory turnover. As a result, they could achieve low levels of inventory risk and associated costs. Walmart's significantly low costs and high levels of efficiency were major factors that drove productivity of the merchandise retail sector over the period of 1995 — 2000 according to a 2001 McKinsey Global Productivity Report.

Figure 5-1

It has given birth to a new industry.

Historically, data was used as an ancillary to core business and was gathered for specific purposes. Retailers recorded sales for accounting. Manufacturers recorded raw materials for quality management. The number of mouse clicks on advertising banners was collected for calculating advertisement revenue. But as the demand for Big Data analytics emerged, data no longer serves only its initial purpose. Companies able to access huge amounts of data possess a valuable asset that when combined with the ability to analyze it, has created a whole new industry.

ITA Software[4] is a private company that gathers flight price

New Words and Expression

gross margin
总利润，（销货）毛利

merchandise/ˈmɜːtʃəndaɪz/ *n*
商品；货物；买卖；销售；经

ancillary/ænˈsɪləri/ *adj*.
辅助的；补充的；附加的

data from almost all major carriers with the exception of Jet Blue and Southwest that sells that information to travel agents and websites.Google acquired ITA in 2011 for $700 million. With Google's expert analytics and more extensive data for processing, ITA today can provide predictions for prices for flights, hotels, shopping and more.

The success of companies like ITA has helped accelerate the boom of Big Data startups. According to the website angel.com, there have been 2,924 Big Data startups from November 2010 to the present. These companies often operate as data analytics companies, data providers or traders, are attracting a lot of attention from investors. In the second quarter of 2015, U.S. venture capital funding of Big Data startups reached $19.19 billion.

Another remarkable case in this emerging industry is last year's strategic partnership between IBM and Twitter (Figure 5-2). IBM and Twitter have partnered up for the purpose of selling analytical information to corporate clients. IBM analyzes Twitter data combined with other public and business sources, "helping businesses tap into billions of real-time conversations to make smarter decisions" according to Glenn Finch, Global Leader Data & Analytics, GBS, The partnership has helped the two companies leverage their respective areas of expertise; IBM with their analytical skills and Twitter for their data.

Figure 5-2

It improves business regardless of company size.

It is obvious that big companies have advantages over smaller

New Words and Expressions

emerging industry
新兴产业

venture capital
<美>风险资本

strategic partnership
战略伙伴关系

ones. By the word "big", I mean companies that generate an enormous amount of data. Tech giants like Amazon and Google will continue to benefit from the sheer volume of data they generate. Amazon currently has approximately 270 million active users in 185 countries and 16 million listing units. Google has approximately 12 trillion monthly searches, which dominates the internet search engine market to the tune of approximately a 90% market share, including over one billion YouTube users and 500 million Google Plus users.

But that is not the end of the story; Big Data actually helps level the playing field. The breakneck-paced development of technology such as processing chips and data storage have reached a point in which companies can retain and utilize information at very low costs. Even with a limited IT budget, companies can still effectively store data. If there is not enough data available in-house, they can cheaply lease data from "data intermediaries". Companies can also hire outside data analytics firms at affordable rates.

An example of successful application is recruitment company Riviera Partners' process in selecting candidates. They cross reference candidates' profiles in their database with public sources to cherry-pick the most appropriate skills and match them to each position. Another example is a restaurant chain that "was able to eliminate the need to live answer handle 60,000 phone calls to their restaurants, allowing employees to focus on in store customers" according to Michael Bremmer, CEO of Telecomquotes.com.

New projects also benefit from Big Data innovation, as described by Kristina Roth, CEO & Founder of Matisia Consultants, "with big data, businesses can learn to improve faster, better, and at lower costs by learning lessons from each improvement project and incorporating them into the next project."

In fact, Big Data applications are bound only by the human imagination. Businesses such as car manufacturers can improve operational efficiency, hospitals can improve patient services and fast food companies can better manage food deliveries. The list goes on and on. Any business that can successfully apply Big Data creates a competitive advantage.

Notably, successful players in Big Data are recognized well by the market. Companies that utilize Big Data are highly valued by

New Words and Expression

stockpile/ˈstɒkpaɪl/ *n.*
（原料，食品等的）储备，储
大量储备

tech giant
科技巨头

dominate/ˈdɒmɪneɪt/ *v.*
支配，影响；占有优势

data intermediary
数据中介

recruitment/rɪˈkruːtmənt/ *n.*
征募新兵；补充；募集

cherry-pick
最佳选择；优选

investors. Companies engaged in Big Data business have relatively high multiples. Investors may not only value their growth but also their intangible assets[5], such as data volume and analytical skills.

Big Data is making a huge impact and will continue to do so as a key driving factor in business performance in years to come.

> ***New Words and Expressions***
>
> **intangible**/ɪnˈtændʒəbəl/ *adj.* 触不到的；难以理解的；无法确定的；<商>(指企业资产)无形的
>
> **intangible asset** 无形资产

Terms

1. Data-directed decision making

Decision-making can be regarded as a problem-solving activity terminated by a solution deemed to be satisfactory. It is therefore a process which can be more or less rational or irrational and can be based on explicit or tacit knowledge.

Human performance with regard to decisions has been the subject of active research from several perspectives:

- Psychological — examining individual decisions in the context of a set of needs, preferences and values the individual has or seeks.
- Cognitive — the decision-making process regarded as a continuous process integrated in the interaction with the environment.
- Normative — the analysis of individual decisions concerned with the logic of decision-making, or communicative rationality, and the invariant choice it leads to.

A major part of decision-making involves the analysis of a finite set of alternatives described in terms of evaluative criteria. Then the task might be to rank these alternatives in terms of how attractive they are to the decision-maker(s) when all the criteria are considered simultaneously. This area of decision-making, although very old, has attracted the interest of many researchers and practitioners and is still highly debated as there are many MCDA methods which may yield very different results when they are applied on exactly the same data. This leads to the formulation of a decision-making paradox.

数据导向决策

决策可以被看作是解决问题的一种行动过程，反复尝试该过程直到得到令人满意的策略终止。因此，这是一个多多少少掺杂理性或非理性的过程，可以基于显性或隐性知识。

对于人类决策的表现，主要从以下几个主题进行研究：

- 心理学——在个人拥有或寻求的一系列需求、偏好和价值观的背景下审视个人决策。
- 认知——决策过程被认为是与环境相互作用并结合的连续过程。
- 规范性——与决策逻辑或交际理性相关的个别决策的分析以及导致的不变的选择。

决策的主要部分涉及根据评估标准描述的一组有限选择的分析。那么决策的任务可能就是在同时考虑所有标准的情况下，考量对决策者的吸引力是怎样的。这个决策领域虽然很老，但吸引了许多研究人员和从业者的兴趣，并且仍然受到高度争议，因为有许多多准则决策分析（Multi Criteria Decision Analysis，MCDA）方法在应用完全相同的数据时可能产生非常不同的结果，这就可能导致制定一个决策时会出现悖论。

2. Statistical model

A statistical model is a class of mathematical model, which embodies a set of assumptions concerning the generation of some sample data, and similar data from a larger population. A statistical model represents, often in considerably idealized form, the data-generating process.

The assumptions embodied by a statistical model describe a set of probability distributions, some of which are assumed to adequately approximate the distribution from which a particular data set is sampled. The probability distributions inherent in statistical models are what distinguish statistical models from other, non-statistical, mathematical models.

A statistical model is a special class of mathematical model. What distinguishes a statistical model from other mathematical models is that a statistical model is non-deterministic. Thus, in a statistical model specified via mathematical equations, some of the variables do not have specific values, but instead have probability distributions; i.e. some of the variables are stochastic.

There are three purposes for a statistical model, according to Konishi & Kitagawa.

- Predictions
- Extraction of information
- Description of stochastic structures

统计模型

统计模型是一类数学模型，它体现了一组关于生成样本数据（以及来自较大群体的类似数据）的假设。统计模型通常以相当理想化的形式表示数据的生成过程。

由统计模型体现的假设描述了一组概率分布，其中一些被假定为是从充分近似特定数据集中提取的抽样分布。统计模型固有的概率分布将统计模型与其他非统计学数学模型区分开。

统计模型是一类特殊的数学模型。统计模型与其他数学模型的区别在于统计模型是非确定性的。因此，在通过数学方程式指定的统计模型中，一些变量没有特定值，而是具有概率分布，即一些变量是随机的。

根据 Konishi＆Kitagawa 的理论，统计模型有三个目的：

- 预测；
- 提取信息；
- 随机结构描述。

3. Competitive advantage

When a firm sustains profits that exceed the average for its industry, the firm is said to possess a competitive advantage over its rivals. The goal of much of business strategy is to achieve a sustainable competitive advantage.

Michael Porter identified two basic types of competitive advantage(Figure 5-3):

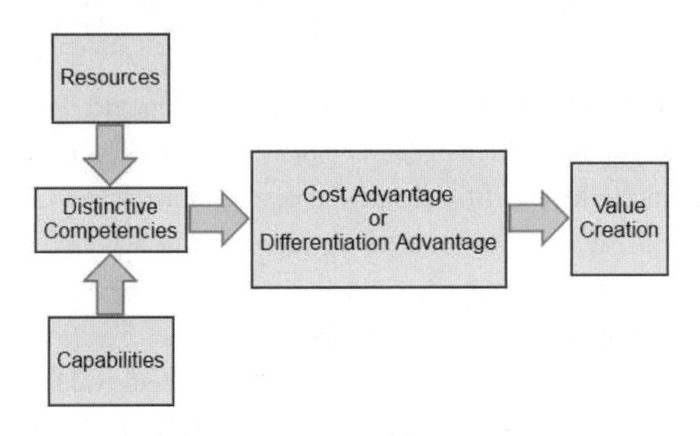

Figure 5-3

- cost advantage
- differentiation advantage

A competitive advantage exists when the firm is able to deliver the same benefits as competitors but at a lower cost (cost advantage), or deliver benefits that exceed those of competing products (differentiation advantage). Thus, a competitive advantage enables the firm to create superior value for its customers and superior profits for itself.

Competitive advantage is a business concept that describes the attribute of allowing an organization to outperform its competitors. These attributes may include access to natural resources, such as high-grade ores or a low-cost power source, highly skilled labor, geographic location, high entry barriers, etc. Access to new technology can also be considered as an attribute of competitive advantage.

竞争优势

当公司维持超过其行业平均水平的利润时，该公司就会被认为比竞争对手具有竞争优势。大多数业务战略的目标是实现可持续的竞争优势。

迈克尔·波特确定了两种基本类型的竞争优势（如图 5-3 所示）：成本优势和差异化优势。

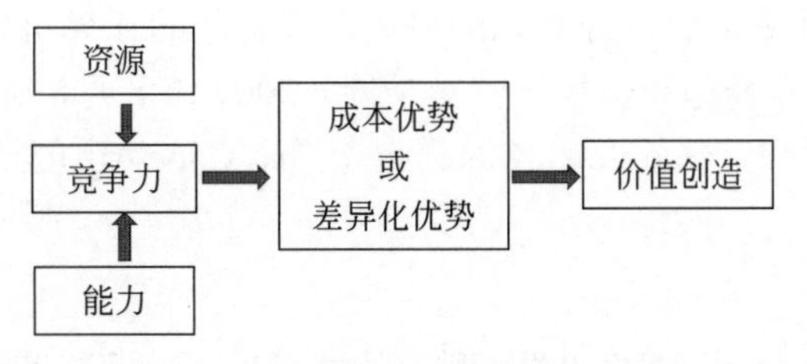

图 5-3

当企业能够以较低的成本（成本优势）就能获得与竞争对手相同的利润，或者提供超过对手竞争产品（差异化优势）的品质时，就存在竞争优势。因此，竞争优势使企业能够为客户创造优越的价值，为自身创造优势。

竞争优势是描述允许组织优于其竞争对手的属性的业务概念。这些属性可能包括获取自然资源，例如高档矿石或低成本电源、高技能劳动力、地理位置、高进入门槛等。获得新技术也可以被视为竞争优势的属性。

4. ITA Software

ITA Software is a travel industry software division of Google, formerly an independent company, in Cambridge, Massachusetts. The company was founded by Jeremy Wertheimer, a computer scientist from the MIT Artificial Intelligence Laboratory and Cooper Union, with his partner Richard Aiken in 1996. On July 1, 2010 ITA agreed to be acquired by Google. On April 8, 2011, the US Department of Justice approved the buyout. As part of the agreement, Google must license ITA software to other websites for five years.

ITA is known for using programming puzzles to attract and evaluate potential employees since 2001. Some of these puzzles have appeared in ads on Boston's MBTA subway system. ITA is also one of the highest-profile companies to base their software on Common Lisp.

In January 2006, ITA received $100 million in venture capital money from a syndicate of five investment firms led by Battery Ventures, marking the largest investment in a software firm in New England in five years. In September 2006, ITA announced a several million dollar deal with Air Canada to develop a new computer reservations system to power its reservations, inventory control, seat availability, check-in, and airport operations. In August 2009, Air Canada announced that the project had been suspended. On March 1, 2012, Google and Cape Air announced that Cape Air had migrated to ITA Software's passenger reservations system.

ITA Software 是 Google 的旅游行业软件部门，原来是位于马萨诸塞州剑桥的独立公司。该公司由麻省理工学院人工智能实验室和库珀联盟的计算机科学家杰里米·韦特海默（Jeremy Wertheimer）和他的合伙人理查德·艾肯（Richard Aiken）于 1996 年创立。2010 年 7 月 1 日，ITA 同意被 Google 收购。 2011 年 4 月 8 日，美国司法部批准收购。作为协议的一部分，Google 必须在此后的 5 年中向其他网站许可使用 ITA 软件。

2001 年来，ITA 以使用编程谜题吸引和评估潜在员工而闻名。其中一些难题出现在波士顿 MBTA 地铁系统的广告中。ITA 也是基于 Common Lisp 的最高级别公司之一。

2006 年 1 月，ITA 从 Battery Ventures 领导的五家投资公司组成的财团获得了 1 亿美元的风险投资资金，这是近五年在新英格兰对一家软件公司的最大投资。 2006 年 9 月，ITA 宣布与加拿大航空公司签署数百万美元的协议，开发新的计算机预订系统，以提供预订、库存控制、座位可用性、登记和机场运营等功能。2009 年 8 月，加拿大航空公司宣布该项目暂停。2012 年 3 月 1 日，Google 和 Cape Air 宣布，Cape Air 已经迁移到 ITA Software 的乘客预订系统。

5. Intangible asset

An intangible asset is an asset that lacks physical substance and usually is very hard to evaluate. It includes patents, copyrights, franchises, goodwill, trademarks, trade names, the general interpretation also includes software and other intangible computer based assets. Contrary to other assets, they generally — though not necessarily — suffer from typical market failures of non-rivalry and non-excludability.

Intangible assets have been argued to be one possible contributor to the disparity between company value as per their accounting records, and company value as per their market capitalization. A number of attempts have been made to define intangible assets:

- Prior to 2005 the Australian Accounting Standards Board issued the Statement of Accounting Concepts number 4 (SAC 4). This statement did not provide a formal definition of an intangible asset but did provide that tangibility was not an essential characteristic of asset.
- International Accounting Standards Board standard 38 (IAS 38) defines an intangible asset as: "an identifiable non-monetary asset without physical substance."

The Financial Accounting Standards Board Accounting Standard Codification 350 (ASC 350) defines an intangible asset as an asset, other than a financial asset, that lacks physical substance.

无形资产

无形资产是一种没有物质形态的资产，通常难以评估。它包括专利、版权、特许经营、商誉、商标、商品名称，通常也包括软件和其他无形资产。与其他资产相反，它们通常不会遭受非竞争和非排他性的典型市场失灵。

考虑到无形资产在公司价值与会计记录价值之间存在差距导致公司价值与市场资本化之间存在差距，人们尝试给出对无形资产的定义。

- 2005年之前，澳大利亚会计准则委员会发布了第4号会计概念表（SAC 4）。该声明没有提供无形资产的正式定义，但确实规定了有形资产不是资产的必要特征。
- 国际会计准则委员会第38号（国际会计准则第38号）将无形资产定义为"无物理实质可辨认的非货币性资产"。

财务会计准则委员会会计准则编制350（ASC 350）将无形资产定义为"缺乏实质的资产（金融资产除外）"。

Comprehension

Blank Filling

1. Proper use of big data goes beyond collecting and analyzing large quantities of data; it also requires understanding ______________ to use the data in______________.
2. Big data analytics can positively impact: product development, ________development, operational __________, customer experience and _________, market demand __________.

Content Questions

1. What is the biggest value created by insights from large data sets?
2. In what ways has Big Data been impacting companies?

Answers

Blank Filling

1. how and when; making crucial decisions
2. market; efficiency; loyalty; predictions

Content Questions

1. The biggest value created by these timely, meaningful insights from large data sets is often the effective enterprise decision-making that the insights enable.
2. It has revolutionized old-school industries. It has given birth to a new industry. It improves business regardless of company size.

参考译文

企业正在利用大数据提供的强大的洞察力来实时确定谁在何时何地做了什么。这些来自大型数据集的实时的、有意义的结论创造出的价值，经常为具有远见的高效企业提供决策。

大数据提供机会的同时也伴随巨大的挑战。掌握大数据管理这门新兴学科的企业可以获得巨大回报，并与竞争对手拉开距离。事实上，美国麻省理工学院斯隆管理学院经济学家 Erik Brynjolfsson 所做的研究表明，使用“数据导向决策”的公司的生产力提高了 5%～6%。大数据的正确使用超出了收集和分析大量数据的范畴，它还需要了解何时和如何使用数据做出决策。

利用正确的数据可以大大提高竞争优势。根据麦肯锡的研究报告，美国医疗保健行业数据的潜在价值可能每年超过 3000 亿美元，其中三分之二将可使国民医疗支出降低约 8%。

当数据管理过程与企业战略相一致时，可以实现财务效益，这时候可能就需要高层管理人员参与确定方向并监督重大决策。

大数据分析可以在以下方面产生积极影响：

- 产品开发；
- 市场发展；
- 运营效率；
- 客户体验和忠诚度；
- 市场需求预测。

大数据是当今商人的流行语。无论行业或公司规模如何，它都会被热议。大数据至少有三种影响公司发展的方式。

1. 它彻底改变了行业的旧模式

9Lenses 公司首席执行官 Edwin Miller 说：“大数据对于从客户关系到供应链管理的业务都产生了巨大的影响，并将持续这样影响下去。”虽然许多人可能没有意识到这一点，但大数据确实对许多传统业务产生了重大影响。

Capital One 公司就是一个很好的例子。20 世纪 90 年代，信用卡行业利用统一定价模式对每个客户收取相同的价格，除了 Capital One。该公司使用基于公共信贷和人口统计数据的统计模型为客户提供“量身定制”的产品。这一创新是 1994—2003 年期间的净收入实现 32%复合年增长的发展基石之一。因此，许多银行已将重点转移到大数据分析，但开拓者似乎保持了优势。2009—2014 年，与美国顶级银行（如花旗集团 11%、美国银行 11%、美国摩根大通 6%）相比，其年度净收入增长了 17%。

劳斯莱斯在应用大数据分析方面的成功影响了飞机发动机制造业。该公司始终如一地监控着约3700台发动机，每台发动机都安装有数百个传感器，用来预测发生故障的时间和位置。劳斯莱斯已经从仅销售发动机转变为销售发动机和监控服务的组合，然后根据使用、维修和更换情况向客户收费。该服务目前占飞机发动机部门年收入的70%以上。

另一个完美的例子是世界上最大的零售商沃尔玛。沃尔玛是当今大数据分析的知名用户，但在20世纪90年代，沃尔玛对零售业进行了革命，它通过一个被称为零售链接（Retail Link）的系统，记录了所有产品的各种属性。该系统为供应商管理自己的产品提供了一种方式，允许他们监控其数据，包括销售和库存量、库存百分比、毛利率和库存周转量。因此可以实现将库存风险和相关成本降低。根据2001年麦肯锡全球生产力报告，沃尔玛的成本显著降低、效率水平高，是促成1995—2000年商品零售行业生产力的主要因素。

2．大数据促使一个新行业的诞生

历史上，数据被用作核心业务的辅助部件，并被收集用于特定目的。零售商记录销售情况以进行会计核算，制造商记录了原材料以进行质量管理，人们收集了广告横幅上的鼠标点击次数以计算广告收入。但随着对大数据分析需求的出现，数据不再仅拥有其作为原始数据时的功能与目的。能够获取大量数据的公司具有宝贵的资产，当与分析能力相结合时，创造了一个全新的行业。

ITA Software是一家私人公司，该公司从几乎所有主要航空公司收集航班价格数据，除了将这些信息出售给旅行社和网站的Jet Blue和西南航空公司。Google于2011年以7亿美元收购ITA Software。通过Google的专家分析和更多的数据处理，ITA今天可以提供航班、酒店、购物等的价格预测。

像ITA这样公司的成功有助于加速大数据创业公司的蓬勃发展。根据网站angel.com提供的数据显示，2010年11月以来，已有2924个大数据创业公司。这些公司经常作为数据分析公司、数据提供商或交易商，受到投资者的关注。2015年第二季度，美国大数据创业公司的风险投资资金达191.9亿美元。

这个新兴行业的另一个显著的例子是去年IBM和Twitter之间的战略合作伙伴关系。IBM和Twitter的合作，目的是向企业客户销售分析信息。全球领先的数据和分析公司GBS的Glenn Finch表示，IBM将Twitter数据与其他公共和商业资源相结合，“帮助企业进行数十亿次实时对话，做出更明智的决策”，该合作伙伴关系帮助两家公司利用各自的专业领域：IBM的分析技能和Twitter的数据。

3．无论公司规模如何，大数据都可以改善业务

很明显，大公司比较小的公司具有优势。“大”公司意味着产生大量的数据。像亚马逊和Google这样的科技巨头将继续受益于它们生成的大量数据。亚马逊目前在185个国家拥有大约2.7亿的活跃用户和1600万个产品名录。Google每月有约12万亿次搜索量，占互联网搜索引擎市场的90%左右，其中包括超过10亿YouTube用户和5亿Google Plus用户。

但这不是故事的终点，大数据实际上有助于平衡竞争环境。处理芯片和数据存储等技术的迅猛发展已经达到了企业以非常低的成本存储和利用数据的一个关键点。即使IT预算有限，公司仍然可以有效地存储数据。如果内部数据不足，也可以从“数据中介”中廉价租用数据。公司也可以以合理的价格聘请外部数据分析公司。

招聘公司Riviera Partners是在招聘候选人方面取得成功的例子。他们在公共资料库的数据库中交叉参考候选人的资料，以挑选最适合的技能，并将其与每个职位相匹配。Telecomquotes.com 首席执行官迈克尔·布雷默（Michael Bremmer）表示，另一个例子是“连锁餐厅”，它消除了对实时处理6万个打给他们餐厅的电话的需求，这样允许员工专注于店内客户。

新的项目也受益于大数据创新，马里西亚咨询公司首席执行官克里斯蒂娜·罗斯（Kristina Roth）介绍说，“通过大量数据，企业可以通过从每个改进项目中吸取教训并加以融合运用到下一个项目，从而更快、更好并降低成本。”

事实上，大数据应用程序只受人的想象力的束缚。汽车制造商等企业可以提高运营效率，医院可以改善患者服务，快餐公司可以更好地管理食品交付，这样继续下去，任何擅长应用大数据的业务都将创造竞争优势。

很明显，擅长大数据的人才得到市场认可。利用大数据的公司受到投资者的高度重视。从事大数据业务的公司的市盈率相对较高。投资者不仅可以评估其增长值，还可以评估其无形资产，如数据量和分析能力。

大数据正在产生巨大的影响，并将在未来几年继续成为业务绩效的主要推动因素。

Text B

Big data is a great resource for driving smart business decisions and changes. Here are the ways that the use of big data is improving how business gets done.

When business leaders hear the term big data, they most naturally think of the massive volumes of data available today. This data is created by e-commerce and omnichannel marketing systems, or IoT-connected devices, or business applications that generate ever more detailed information about transactions and activities. And those are just a few examples.

The sheer scale of the data is daunting, maybe even overwhelming in some cases. But there are great business benefits to be gained by analyzing sets of big data. We'll explore some of these benefits below. Now, let's look at the ways in which big data can improve the way we do business.

Better customer insight

When a modern business turns to data to understand its customers — whether individually or in categories — it has a wide range of sources to choose from. Big data sources that shed light on customers include the following:

- Traditional sources of customer data, such as purchases and support calls;

New Words and Expressions

omnichannel/ˈɒmnɪˈtʃæn(ə)l/ *n.* 全渠道

sheer/ʃɪə(r)/ *adj.* 程度深的，数量大的

daunting/ˈdɔːntɪŋ/ *adj.* 使人畏惧的，使人气馁的

- External sources, such as financial transactions and credit reports;
- Social media activity;
- Data from internal and external surveys; and computer cookies.

Clickstream analysis of e-commerce activity is especially useful in an increasingly digital marketplace, shedding light on how customers navigate through a company's various webpages and menus to find products and services. Companies can see which items customers added to their carts but perhaps removed or later abandoned without purchasing; this provides important clues as to what customers might like to buy, even if they don't make a purchase.

Not only online stores, but brick-and-mortar locations can also glean useful understanding of their customers, often by analyzing video to learn how visitors navigate through a physical store compared with their navigation of a website.

Increased market intelligence

Just as big data can help us analyze the complex shopping behavior of customers in more detail, it can also deepen and broaden our understanding of market dynamics.

Social media is a common source of market intelligence for product categories ranging from breakfast cereal to vacation packages. For almost any commercial transaction you can imagine, there are people out there sharing their preferences, their experiences, their recommendations ... and their selfies! Yes, even of their breakfast fare. These shared opinions are invaluable for marketers.

In addition to competitive analysis, big data can also help in product development: by prioritizing different customer preferences, for example.

In fact, big data does not just assist with modern market intelligence; in almost any e-commerce or online market, almost all market intelligence is driven by diverse, ever-changing data.

Smarter recommendations and audience targeting

In our lives as consumers, we are now so familiar with recommendation engines that we might not be aware of how much they have evolved since the advent of big data. At one time, the predictive analysis for recommendation engines was quite simple:

New Words and Expressions

clickstream/ˈklɪkstriːm/ *n.* 点击流量

brick-and-mortar /braikəndˈmɔːtə/ *adj.* 实体的

transaction/trænˈzækʃ(ə)n/ *n.* 交易，业务，买卖

prioritize/praɪˈɒrətaɪz/ *v.* 按优先顺序列出；优先考虑（处理）

advent/ˈædvent/ *n.* 出现，到来

association rules which found those common items in market baskets. You can still expect to find this as a feature on e-commerce websites telling us that customers who bought widgets also bought fidgets.

Newer recommendation systems are much smarter than that, building on the sophisticated customer insights we have already discussed, with the result that they can be more sensitive to demographics and customer behavior. These systems aren't limited to e-commerce, either. A friendly waiter's recommendations may well be data-driven — decisions prompted by a point-of-sale system that evaluates stock levels in the pantry, popular combos, high-profit items and even social media trends. When you share a picture of your meal, you are providing yet more input for the big data engines to digest.

Streaming content providers use even more sophisticated techniques. They may not even ask customers what they want to see next: Even before the current movie, program or song finishes, the next selection fades in, keeping viewers binge-watching by utilizing their own preferences combined with a great deal of big data analysis gleaned from other users and social media.

New Words and Expressions

widget/ˈwɪdʒɪt/ *n.*
装饰物；小部件

fidget/ˈfɪdʒɪt/ *n.*
烦躁

demographics/ˌdeməˈgræfɪks/
人口统计

pantry/ˈpæntri/ *n.*
食品储藏室

binge-watching/bɪndʒ wɑːtʃɪŋ
adv. 无节制地

glean/gliːn/ *vt.*
收集（资料）

Note:

The text is adapted from the website:

https://www.techtarget.com/searchbusinessanalytics/feature/6-big-data-benefits-for-businesses

参考译文

大数据推动了商业的智能决策和变革，以下是利用大数据改善业务的方法。

当企业领导听到大数据这个词时，他们会最自然地想到可供使用的海量的数据。这些数据来自电子商务和全渠道营销系统、物联网连接的设备、详细的交易信息，而这些仅仅是数据来源的一部分。

数据的庞大规模令人望而生畏，在某些情况下甚至可能让人不知所措。但是，对大数据的分析，可以使企业获得巨大的利益，我们将探讨大数据的一些好处。现在，让我们来看看大数据是如何改善企业业务的。

更好地洞察客户

当现代企业利用数据来了解客户时，无论是了解客户个人还是某类客户，数据的来源都非常广泛，揭示客户情况的大数据来源包括：

- 传统的客户数据来源，如购买信息和客服电话数据。
- 外部来源，如金融交易数据和信用报告。
- 社交媒体活动。

- 来自内部和外部调查的数据；以及计算机存储在用户本地终端上的数据。

在越来越数字化的市场中，点击流量分析的作用也日益凸显。通过点击流量分析可以了解客户如何通过公司的各种网页和菜单来寻找产品和服务，同时也能看到客户添加到购物车却没有购买的商品，这可以帮助商家更好地判断客户的喜好。

不仅是网上商店，实体店也可以通过视频来观察顾客的购物路线，从而像网店导航页那样更好地了解客户需求。

洞察市场

正如大数据可以帮助我们详细地分析顾客的购物行为一样，它也可以加深和扩大我们对市场动态的了解。

社交媒体是市场数据的一个常见来源：从早餐的麦片到度假的套餐类别，几乎你能想象到的都有人在社交媒体分享。人们愿意在社交媒体分享他们的偏好、经验、建议……以及他们的自拍！是的，甚至包括他们的早餐。这些分享对营销人员来说是非常宝贵的。

除了竞争分析，大数据还可以通过优先考虑不同客户的偏好来开发产品。

事实上，大数据不仅仅能够帮助洞察市场；在几乎所有的电子商务或在线市场，几乎所有的市场情报都是由多样化的、不断变化的数据驱动的。

智能推荐和受众定位

作为消费者，我们对推荐引擎已经习以为常，以至于我们可能都没有意识到，自从大数据出现以来，推荐引擎已经发展到了何种程度。以前推荐引擎的预测分析是相当简单的：在市场里找出与那些常见物品相关联的商品。你现在仍然能看到网站上的这个功能，它可以告诉你购买了某个产品的客户同时也会购买什么其他产品。

新的推荐系统比这要智能得多，除了我们已经讨论过的对客户的洞察，它对人口统计学和客户行为更加敏锐。这些系统也不仅局限于电子商务，一个友好的服务员的建议很可能也是由数据驱动的：销售系统会分析顾客所购商品的剩余数量、受欢迎的商品组合、利润高的商品，甚至是顾客的社交媒体。当你分享你的饭菜照片时，你都是在为大数据提供可用的信息。

流媒体内容供应商使用的技术更为复杂，他们通过自己的喜好，并结合从其他用户和社交媒体收集到的数据分析，甚至可以不问顾客接下来想看什么，在当前的电影、节目或歌曲结束之前，就会弹出下一个选择，供观众持续欣赏。

Chapter 6

Application of Big Data

Text A

Since Big Data is the buzzword today, and it is like the world rushing behind this newly born celebrity, you must be wondering where exactly does this concept fit in? Without any introduction on the much-known technology now, we tell you the real-life applications that would have been strangled to death had big data had not been around.

- Customer Analysis — Here, according to their past choices and dislikes, companies use big data to understand their customers and target them accordingly. Ever wonder how Google displays the ad. for the footwear you once viewed on your online shopping site? That is because every click of yours is a data crucial to the website, which is tracked to treat you likewise.
- Optimizing Business processes — Based on predictions from social media data and other trends from which useful information is chunked out, retailers optimize their stocks. Amazon is soon going to launch its delivery drone, which will make us of Big Data to get live traffic information from its route, and find the shortest possible path.
- Performance Optimization — For the individual self, it can be used to effectively analyze data from wearable gadgets[1] like Fitbits, smart watches, which track the physical activity of a user and suggest on improving the fitness accordingly.

New Words and Expressions

optimize/ˈɒptɪmaɪz/ *v.*
优化；完善

wearable gadgets
可穿戴设备

- Healthcare sector — DNA strings can be decoded and patterns of diseases can be predicted within minutes, thanks to Big Data! Clinical trials in the future would not restrict themselves to a sample data size but can include everyone. Monitoring of epidemics too has been possible due to this technology.
- In Security and law Enforcement — The NSA[2] in US uses big data extensively to foil terrorist plots. It can even be used to predict criminal activities and cyber security breaches.
- In the making of a smart city — While there is so much to hear about smart cities, what is going into the making of them is essentially big data. Smart traffic systems will be possible by real-time analysis of data from many sensors inside vehicles and streets. Internet of things is another such concept which will help almost everything to be connected to the Internet. And, that means everything producing humongous data which needs storage and analysis and curtain.

It is not only these fields where Big Data is playing a big role, but many other such applications where it proudly finds its place (shown in Figure 6-1). It is not imperative, but it is a necessity now that we have conceived the idea and one so far with it. It has the potential to change the world, for the better, of course!

Five amazing "real-world" uses of big data

As an Information and Data Management professional, you've probably been asked what you do for work and found yourself trying to describe Big Data. Even those of us in the industry struggle sometimes to adequately define the breadth and scope of what Big Data is, and what it can do.

Or maybe you've been considering your next career move, and what the options are for using your skills and experience in Big Data? It can be a challenge to see the forest for the trees when we get bogged down in one project or job role, and lose sight of the bigger picture and what other opportunities might exist. With these thoughts in mind, here are what we consider to be some of the most exciting and innovative real world applications for Big Data today (shown in Figure 6-1).

New Words and Expressions

clinical/ˈklɪnɪkəl/ *adj.*
临床的

epidemic/ˌepɪˈdemɪk/ *n.*
（疾病的）流行，传染

sensor/ˈsensər/ *n.*
传感器

Security and law Enforcement
安全执法

foil/fɔɪl/ *v.*
挫败，阻止，制止

terrorist plots
恐怖分子的阴谋

humongous/hjuːˈmʌŋgəs/ *adj.*
极大的，硕大无比的

curation/kjʊəˈreɪʃən/ *n.*
策展；治愈，治疗

imperative/ɪmˈperətɪv/ *adj.*
极重要的

scope/skəʊp/ *n.*
范围

get bogged down in
陷入泥沼

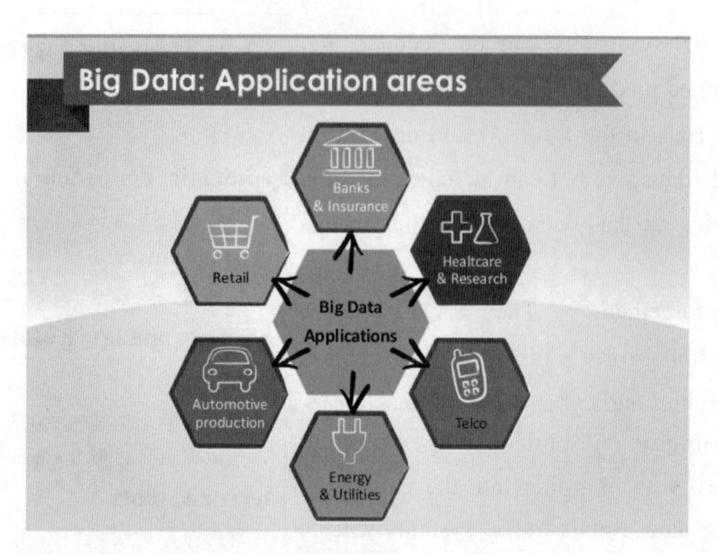

Figure 6-1

1. Disaster relief

Simply put, Big Data is the gathering of tremendous amounts of data from multiple sources, and analyzing it for insights and solutions. In the case of disaster management, it can make sense of chaos in ways that can literally save lives. Crowdsourced, grassroots efforts have arisen in response to earthquakes, typhoons and earthquakes, helping victims and relief workers in equal measure. And increasingly, officials at humanitarian organizations and government agencies are formalizing efforts to fund and maintain networks of maps, satellite images, communications and infrastructure data, and other information that will bolster the speed and efficiency of their efforts. Whether prevention or planning, relief or recovery — Big Data can play a crucial role in improving the way we respond to disaster.

2. Public health and research

Another sector where Big Data is improving lives is public health. Through the collection and analysis of large datasets, Big Data specialists are making astonishing strides in genetic and medical research, and creating improved outcomes for the treatment and prevention of disease — not to mention increasing value for the healthcare system through coordinated care initiatives. To give an example, the top public healthcare system in Germany implemented a proprietary analytics tool that mines and predicts outcomes for data that is growing exponentially. By targeting their

New Words and Expressions

typhoon/taɪˈfuːn/ *n*. 台风

bolster/ˈbəʊlstər/ *v*. 支撑；加固

proprietary/prəˈpraɪətəri/ *adj* 专有的

exponentially/ˈekspəˈnenʃəlɪ/ *adv*. 以指数方式

research, they were able to identify patients with a 90% chance of hospitalization within the year with a congestive heart failure diagnosis. On an even larger scale, Big Data can similarly be used to monitor and predict epidemics, and hopefully prevent them.

3. Weather forecasting and climate change

Thanks to the growth in both the numbers of sensors and satellites in operation, and in the speed at which the input can be processed, Big Data plays an important role in predicting weather patterns and in particular, cataclysmic events. Data improves the ability to accurately predict the timing and intensity of storms, potentially saving human lives and minimizing destruction of property and infrastructure. Taking the longer view, data also increases the understanding of the impact of climate change, both from a meteorological and economic perspective, and is pivotal in the work being done in natural resource management, food and agriculture, ecology and material sciences — the list goes on. Recognizing this, last year the U.N.'s Global Pulse program (for the use of data in development and humanitarian efforts) launched its Big Data Climate Challenge.

4. Financial services

Weather isn't the only thing that calls for an accurate forecast. Financial institutions are increasingly harnessing the power of Big Data for sophisticated financial modeling, determining demand and costs, and cushioning the impact of financial and currency upheavals. Private corporations, such as banks and brokerages, as well as governing agencies and non-profits, are working to predict the occurrences and effects of global financial events on both a micro and macro scale. By creating detailed, scenario-based forecasts, organizations can identify key weaknesses for global economies and financial markets under various conditions and develop courses of action. Like retailers before them, financial institutions are also more reliant on data to obtain a 360-degree view of their customers, to better target products and services and create a competitive advantage.

5. Sport and entertainment

Fortunately, not every application of data has life-or-death implications (although arguably that could depend on how ardently you support your team). The world of sport relies on the analysts in its lineup just as much as its top goal-scorers. From daily decisions

New Words and Expressions

hospitalization /hɒspɪtəlaɪˈzeɪʃən/ *n.*
医院收容，住院治疗

cataclysmic/ˈkætəklɪzəm/ *adj.*
大变动的

meteorological /ˌmiːtiərəˈlɒdʒɪkəl/ *adj.*
气象的

cushion/ˈkʊʃən/ *v.*
对（某事物的影响或力量）起缓冲作用

upheaval/ʌpˈhiːvəl/ *n.*
激变；动荡；剧变

reliant on
依赖

implication/ˌmplɪˈkeɪʃən/ *n.*
含义；暗指，暗示

ardently/ˈɑːdntlɪ/ *adv.*
热心地

such as choosing the starting players, to bigger issues such as developing long-term prospects and creating marketing franchises, Big Data is a star performer. Data also drives our favourite entertainment options, such as Apple Music or Netflix, which use vast amounts of data to optimize streaming performance and personalize recommendations for users.

> ***New Words and Expressions***
>
> **franchise**/ˈfræntʃaɪz/ *n.* 特许经销权

Note:

The text is adapted from the website:

http://www.linkedin.com/pulse/awesome-real-world-applications-big-data-quest-eduventures.

Terms

1. Wearable gadgets

Wearable gadgets are usually high-tech equipment, which are so small that people can wear on heads or wrists.

可穿戴设备通常是高科技设备，这些设备非常小，人们可以戴在头上或手腕上。

2. NSA

National Security Agency，国家安全局。

Comprehension

Blank Filling

1. According to customers' past choices and dislikes, companies use big data to ____________ and ____________.
2. Based on ________ from social media data and other ______ from which useful information is chunked out, retailers optimize their stocks.
3. DNA strings can be ______ and patterns of diseases can be _______ within minutes, thanks to Big Data! Clinical trials in the future would not restrict themselves to a sample data size but can include ________.
4. Smart traffic systems will be possible by real-time analysis of data from many ______ inside vehicles and streets.

Content Questions

1. Which areas can Big data be applied in?
2. What can big data be used for?
3. How the retailers optimize their stocks?
4. Why Big Data can play a crucial role in improving the way we respond to disaster?
5. Why Big Data plays an important role in predicting weather patterns?

Answers

Blank Filling

1. understand their customers; target them accordingly
2. predictions; trends
3. decoded; predicted; everyone
4. sensors

Content Questions

1. Disaster relief, Public health and research, Weather forecasting and climate change, Financial services, Sport and entertainment and so on.
2. Customer Analysis, Optimizing Business processes, Performance Optimization, Healthcare sector, In Security and law Enforcement, In the making of a smart city.
3. Based on predictions from social media data and other trends from which useful information is chunked out.
4. Officials at humanitarian organizations and government agencies are formalizing efforts to fund and maintain networks of maps, satellite images, communications and infrastructure data, and other information that will bolster the speed and efficiency of their efforts.
5. Data improves the ability to accurately predict the timing and intensity of storms, potentially saving human lives and minimising destruction of property and infrastructure. Data also increases the understanding of the impact of climate change, both from a meteorological and economic perspective, and is pivotal in the work being done in natural resource management, food and agriculture, ecology and material sciences.

参考译文

当今时代，大数据是最流行的词汇，它就像一个正当红的明星。你一定想知道大数据都应用在了哪里？可以说，如果没有它，现实生活中的很多应用程序就会无法正常运行。

- 客户分析——根据客户以往表现出来的喜好，公司使用大数据来了解客户并对其进行个体定位。当你在购物网站浏览了鞋类后，Google 就能向你显示关于鞋的广告，这是因为对于网站来说，用户的每一次点击都是收集其喜爱偏好的重要的数据。
- 优化业务流程——基于社交媒体和其他数据的趋势走向，零售商可以优化其库存。亚马逊即将推出无人机，这将使我们在大数据的帮助下，获取实时交通信息，并找到综合最优路径（综合考虑路线长短、道路拥堵等情况）。
- 性能优化——目前，可穿戴式设备（如 Fitbits 系列、智能手表等）可以收集人类个体的数据进行分析，并跟踪用户的身体活动，给出更适合个人的关于运动、保养等的建议。

- 医疗行业——解码DNA链，可以在数分钟内预测疾病的种类，这都要归功于大数据！未来的临床试验不会局限于单一样本（志愿者、患者等）的数据范畴，而是包括所有人（对所有人均有效）。基于此技术，未来对疫情监测也是可行的。
- 安全和执法——美国国家安全局广泛使用大数据来遏制恐怖主义活动的。甚至可以用来预测犯罪活动和网络安全漏洞。
- 打造智能城市——众所周知，智能城市都源自于大数据。通过实时分析车辆和街道中许多传感器的数据，从而实现智能交通系统。物联网也是一个新的概念，也就是物物相连的互联网，这就意味着物联网中所有的“物”都会产生数据，如此海量的数据需要进一步的存储分析和管理。

大数据不仅在这些领域发挥了重要作用，在许多其他类似应用中，也能看到它的影子（如图6-1）。它可能不是势在必行的，但现在我们必须把迄今为止出现的理念构想出来。当然，它有更好的改变世界的潜力！

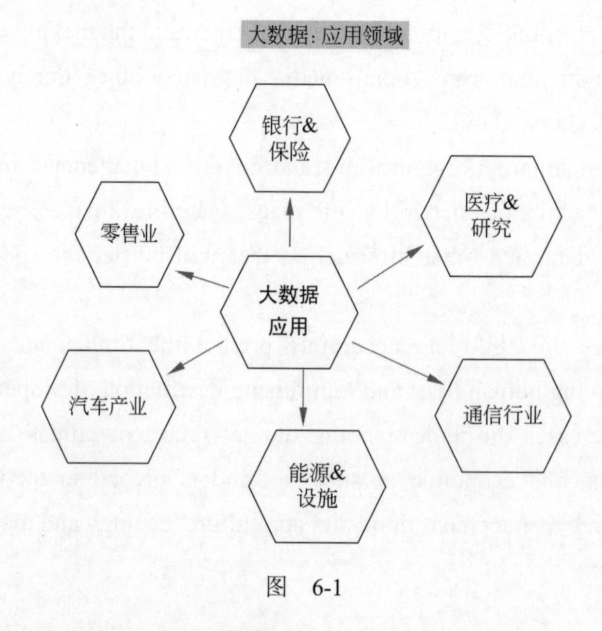

图 6-1

5个惊人的“现实世界”大数据的应用

作为信息和数据管理专业人士，可能经常被询问自己所做的工作，并尝试描述大数据。即使我们这些业内人士，有时也要费心思确定大数据的广度和范围，以及它能做什么。

可能你正在考虑你的下一个职业生涯，利用你的技能和经验在大数据中做着何种选择？当我们在一个项目或工作角色中遇到问题时，很容易一叶障目，从而忽略了更大的图景和其他可能存在的机会。那就带着这些想法，来看看当今现实世界最令人兴奋和最具有创意的大数据应用。

1. 救灾

简单来说，大数据就是从多个来源收集大量数据，分析这些数据并给出解决方案。在救灾中，它可以给救灾人员提供方案。众所周知，它在台风和地震救灾中做出了贡献，帮助灾民和救援人员等。越来越多的人道主义组织和政府机构的官员正在努力资助和维护地图、卫星图像、通信和基础设施数据网络等信息，并改善其工作的速度和效率。无论是预防、规划、救济或恢复，都体现了大数据在改善我们应对灾难的方式方面发挥着关键作用。

2. 公共卫生与研究

大数据也在公共卫生方面改善着生活。通过收集和分析数据，大数据专家在遗传和医学研究方面取得了惊人的进步，并在治疗和预防疾病中创造了更好的成果，通过完善护理措施增加医疗系统的价值。例如，德国最重要的公共医疗系统使用了专有的分析工具，预测数据呈指数级增长的结果。通过他们的研究，可以有 90%的概率识别出一年内因充血性心力衰竭住院的患者。在更大程度上，大数据也可以用于监测和预测流行病，并有希望阻止它们的发生。

3. 预报天气和气候变化

由于传感器和卫星数量的增长以及处理速度的提高，大数据在预测天气，尤其是灾难性事件中起着重要的作用。数据能够准确预测风暴时机和强度，有可能会挽救生命并能最大限度地减少财产和基础设施的破坏。长远来看，数据还增加了对气候变化影响的理解，无论是从气象、经济、自然资源管理、食品和农业、生态和材料科学方面来看，它所做的工作都很关键。认识到这一点，去年联合国的“全球脉动”计划（数据在促进发展和人道主义方面的应用）发起了大数据气候挑战。

4. 金融服务

天气并不是唯一需要精准预报的领域。金融机构越来越多地利用大数据的力量进行复杂的金融建模，以确定需求和成本，缓冲金融和货币动荡的影响。私营企业，如银行和经纪公司，以及管理和研究机构，在微观和宏观尺度上努力预测全球金融事件的发生和影响。通过建立详细的基于情景的预测，企业可以在各种条件下找出全球经济和金融市场的关键弱点，并制定行动计划。与零售商一样，金融机构也更依赖于数据来获得客户的全方位视角，从而更好地瞄准产品和服务，创造竞争优势。

5. 体育和娱乐

幸运的是，并不是每一个大数据应用程序都有生死攸关的影响（尽管可以说这可能取决于你如何支持你的团队）。体育世界需要的分析师数量与其顶级目标得分手的数量一样多。例如，从选择上场球员的日常决策到长期发展的前景并在创造营销专营权这种更大的问题，大数据很在行。大数据还推动着我们最喜欢的娱乐选项，如 Apple Music 或 Netflix，它们使用大量的数据来优化性能和进行用户个性化推荐。

Text B

Interesting Application of Big Data Analytics

Humans are doing two activities continuously: first is breathing and the second one, generating data continuously. There is a constant data generation that takes place. With the advent of mobile technology, this process has escalated manifold giving rise to humongous amount of data. To give you the enormity of data, the Ericsson report said that the world wide data generated through

> ***New Words and Expressions***
>
> **escalate**/ˈeskəleɪt/ *v.* （使）增强；（使）扩大
>
> **enormity**/ɪˈnɔːməti/ *n.* 巨大；严重性；深远影响

mobile reached 7ZB in 2014. Also, the report predicts data generation to reach 2GB per mobile device per month by 2018.

The companies providing communication services (Mobile, telecommunication) can use this data to strategize wide range of activities in order to provide more competitive offers to the consumers, work on the pricing and packages of the product. The data can be used to enhance consumer experience, which further increases the customer loyalty. This can be done by creating smarter networks along with the extension of various functionality in order to provide an organized networked society.

Data analytics is an opportunity for the marketers to get maximum benefits to the organization. The most important usage of big data analytics is to understand the consumer behavior. Using the findings the user can effectively target the right audience thus making the optimum use of the budget. Predictive analytics enhance the targeting even more. For example, a company making baby products will benefit if they come to know which customer is due to deliver a baby in the recent future.

Secondly, big data is also used to optimize various business processes. One particular industry which is using this technique to a great extent is the supply chain industry. It is done by collecting data using various sensors which track the movement of the goods and vehicles. Another industry where the analytics are used extensively is to optimize the talent acquisition process by the HR professionals.

Big data has its applications on individual levels as well. For example, various wearable technologies like smart watch and smart fitness brand generate data at individual level. It uses analytics accompanied with data visualization technique in order to make the finding more engaging and presentable to the user. Another important application of big data is the financial trading. It is an area which involves high frequency of transactions. The brevity of the application is critical in this domain as it includes buying and selling of huge amount of capital in split seconds.

Apart from the above mentioned domains, big data has a big role to play in other fields like health, insurance, research, etc. Gradually, the corporate ecosystem is adapting to this new tool in order to optimize the business bottom-line.

New Words and Expressions

engaging/ɪnˈgeɪdʒɪŋ/ *adj.*
迷人的；吸引人的

presentable/prɪˈzentəbl/ *adj.*
像样的；拿得出的；中看的；中听的

brevity/ˈbrevəti/ *n.*
简洁；短暂

Note:

The text is adapted from the website:

http://www.linkedin.com/pulse/4-interesting-application-big-data-analytics-richa-kapoor.

参考译文

大数据分析的有趣应用

人类在不断地做两个活动：第一是呼吸，第二个是不断生成数据。随着移动技术的出现，这个过程已经升级，导致数据量越来越庞大。爱立信公司的报告指出，为了给人们提供巨大的数据服务，到 2014 年，通过手机生成的全球数据达到 7ZB。此外，该报告预测到 2018 年，每个移动设备每月的数据生成量将达到 2GB。

提供通信服务的公司（手机公司、电信公司）可以根据这些数据来策划很多的活动，用来向消费者提供更具竞争力的优惠，并对产品的定价和包装进行处理。数据可用于提升消费者体验，进一步提升客户忠诚度。这些都可以通过创建更智能的网络以及各种功能的扩展来实现，提供有组织的网络社会。

数据分析是营销人员获得最大收益的机会。大数据分析的最重要用途是了解消费者行为。利用分析结果，公司可以有效地瞄准正确的受众，从而最大限度地利用预算。预测分析可以准确发现目标。例如，一家生产婴儿产品的公司如果知道哪个客户在最近将要生下宝宝，将会由此获益。

其次，大数据也用于优化各种业务流程。在很大程度上使用这种技术的是供应链行业。它们通过使用跟踪货物和车辆运动的各种传感器收集数据来完成这个过程。广泛使用大数据分析的另一个行业是人力资源，优化专业人才的获取过程。

大数据也适用于各个应用层面。例如，智能手表和智能健身品牌的各种可穿戴技术产生大量数据。使用基于数据可视化技术的分析，可以找到更具吸引力和可视的方案给用户。大数据的另一个重要应用是金融交易。这是一个涉及高频率交易的领域。应用程序的简洁性在这个领域至关重要，因为它在几秒钟内就有大量的资本交易发生。

除了上述领域外，大数据在健康、保险、研究等领域发挥着重要作用。企业生态系统正在逐渐适应这种新工具，以优化业务底线。

Chapter 7

Big Data in Recruitment Marketing

Text A

The role of the HR pro is drastically changing, as well as the role of recruiters. Even those smaller, sort of autopilot tasks are evolving with data-driven changes. As recruiters pick up and learn one tool, another, more sophisticated tool comes along to replace it. Well, the recruitment tool of right now is big data. Big data can be utilized in almost any area of business, but right now let's talk about big data and its application in recruitment marketing.

"What's certain is that big data is the future of job recruiting and development, and understanding how to make sense of it will be critical to a company's success. These days, big data is helping fast-growing companies find their perfect engineers, developers and executives." — Michael A. Morell, Riviera Partners

As the role of the recruiter evolves, recruiters are picking up several of the skills that traditionally marketers have required, but they are also becoming quite the data analysts (don't worry, reporting and analysis tools do most of the heavy lifting). Recruiters are now using data-driven recruitment marketing to strategically attract and retain quality talent — both mounting concerns for the majority of business leaders.

No More Pin the Job on the Donkey[1]

The days of post and pray are over. Recruiters can now gather a wealth of actionable data from job boards. Objective information can be collected from job postings to help recruiters use this

New Words and Expressions

sophisticated/səˈfɪstɪkeɪtɪd/ *adj*. 精密的，复杂的；高级的

retain/rɪˈteɪn/ *v*. 保持；保留；保有

mounting/ˈmaʊntɪŋ/ *adj*. 增加的；加剧的

pin/pɪn/ *v*. （用别针等）别住，钉住，固定住

a wealth of 大量的，丰富的

popular recruitment medium optimally, and craft the best recruitment messaging possible. Then, find the magical combination between placement and messaging factors.

Next, recruiters are able to target specific talent sources through this data, to reach the highest likelihood of enlarging the talent pool for each position. Previously, targeting sources was a bit of a risky business. To allocate spend in a niche manner, could mean totally blowing your budget on an attraction avenue with no ROI. Making you look like a… donkey.

Now, applying big data to this process will result in the creation of free-flowing talent pipelines, which in turn will decrease time-to-fill rates. Job boards are no longer the frustrating and costly game of pin the job on the donkey. Recruiters can now use job boards with data-driven action. All of this contributes to the smarter and more effective allocation of your recruitment budget.

The Magic 8 Ball[2] of Recruiting

Until very recently, recruiters have been obsessed with a small pool of recruiting metrics. Well, of course now we know recruiters have to stop looking to the past for answers, and make the big switch to predictive analytics. Predictive analytics are found in the recruiting data that you already have; they are simply the connection of those data points. Predictive analytics help recruiters ask and answer the right questions.

- How many days does it take to hire an IT professional in Los Angeles?
- On what day, during which hours do IT professionals apply for jobs?
- What job boards do IT professionals from Los Angeles use most often?
- What is the application abandonment rate for an IT professional from Los Angeles on a non-mobile optimized site?

I could go on forever with questions that can help recruiters target, attract and sign the talent they need, exactly when they need it. Big data connects the data points that give recruiters the answers. Big data reveals the trends we don't see on the surface (or even under it, for that matter). This knowledge arms recruiters with the confident agility to tailor their methods and budgets for the best recruitment marketing efforts possible.

New Words and Expressions

allocate/ˈæləket/ *v.*
分配；分派

niche manner
利基方式

ROI
投资回报率

agility/əˈdʒɪlətɪ/ *n.*
敏捷，机敏

The big data movement has proven effective and will, therefore, last. A recent Deloitte study revealed 57% of human resources departments increased their spend on analytics. This is the direct result of tools that render previously useless data, now actionable and objective information. Recruiters today have solid data to replace assumptions and best guesses.

As mentioned before, recruiters will not have to be analytical, mathematical clairvoyants, but they will need to be armed with the appropriate reporting and analytics tools. No one can achieve this level of highly targeted recruitment marketing without technology that turns backward-facing data into useful analytics.

Bio: Kelly Robinson

Kelly Robinson is the founder and CEO of Broadbean Technology, a sourcing and recruitment technology company. Broadbean Technology has created a strong global presence with offices in the US, Europe and Australia. The company remains true to the core fundamentals of its inception: "Keep it light and fun while getting the job done!" Kelly writes about leadership and culture, as well as reducing friction in the candidate experience.

The Importance of Big Data in Recruitment

The recruitment space is fast moving and heavily reliant on technology — it's become an industry of first-adopters, first within social media and now within Big Data and analytics.

Big Data has been used in a variety of industries to make better decisions and also to drive new avenues for revenue and growth. It is however not the only tool that companies need in order to recruit the best talent, but it certainly gives companies another option from which to design and implement new hiring strategies.

If a company is experiencing high turnover, it's likely that there will be an entrenched, reoccurring problem that's causing the issue. Whether it be commuting times, a repeated incident with a specific manager or inflexible working hours, data can help find solutions to these problems and it's because this that analytics within HR has a high adoption level.

Big Data has made great strides in a company's ability to predict when and how it should approach potential employees. From understanding which boards will yield the most responses, to discovering a specific time which prospective candidates are more likely to respond to a job role, predictive analytics uses huge pools

New Words and Expressions

human resources
人力资源

clairvoyant/ˌkleəˈvɔɪənt/ *n.*
有洞察力的人

friction/ˈfrɪkʃən/ *n.*
摩擦力；摩擦

reliant/rɪˈlaɪənt/ *adj.*
信赖的，信任的

avenue/ˈævənjuː/ *n.*
途径

turnover/ˈtɜːnˌəʊvər/ *n.*
人员调整率

entrenched/ɪnˈtrentʃt/ *n.*
根深蒂固的

inflexible/ɪnˈfleksəbəl/ *adj.*
不可改变的

of data to discover valuable trends and the facets of a job description which attracts the best talent to the role.

Start-ups such as "Gild" have also caught the eye of many, helping recruiters delve deeper into the actions of web developers and scouring their public code and professional knowledge to give recruiters an accurate image of the quality of their work. Although concentrated within a niche area, there's considerable opportunity for this to be expanded to new roles in the near future.

As mentioned before, technology in itself is having a real impact on the way recruiters attract and retain employees. With software programmes such as "the Resumator" now used to compliment more hard-line data driven programmes, technology is reshaping the recruitment world and allowing companies to attract the best talent.

New Words and Expressions

scour/skaʊər/ *v.*
擦净；刷掉

Note:

The text is adapted from the website:

http://www.linkedin.com/pulse/application-big-data- recruitment-marketing-kelly-robinson;

http://www.linkedin.com/pulse/importance-big-data-recruitment-elliot-pannaman.

Terms

1. Pin the Job on the Donkey　Pin the Tail on the Donkey

Pin the Tail on the Donkey is a game played by groups of children. A picture of a donkey with a missing tail is tacked to a wall within easy reach of children. One at a time, each child is blindfolded and handed a paper "tail" with a push pin or thumbtack poked through it. The blindfolded child is then spun around until he or she is disoriented. The child gropes around and tries to pin the tail on the donkey. The player who pins their tail closest to the target, the donkey's rear, wins.

Idiomatically, the term can be used derisively for any assigned activity which is pointless or for which a person has been handicapped (blindfolded).

在驴上挂尾巴是一群孩子玩的游戏。一头驴尾巴丢失的照片贴在孩子够得到的墙上。每一个孩子都被蒙住眼睛，并拿着一个带推针或图钉的纸“尾巴”。让被蒙住眼睛的孩子旋转，直到他或她迷失方向。孩子们摸索着，试图把尾巴钉在驴子上，谁贴的尾巴最靠近目标谁赢得比赛。

这个词通常用于指任何没有意义的活动或者指一个人被蒙住眼睛后所做的活动。

2. Magic 8-Ball

The Magic 8-Ball is a toy used for fortune-telling or seeking advice, developed in the 1950s and manufactured by Mattel. It is often used in fiction, often for humor related to its giving very accurate, very inaccurate, or otherwise statistically improbable answers.

这个神奇的8球是一个用于算命或寻求建议的玩具，它是在20世纪50年代开发的，由美泰公司制造。它经常被用于小说中，通常是关于它的诙谐表达，有时神奇8球所给答案非常准确，有时非常不准确，或者在统计上不太可能。

Comprehension

Blank Filling

1. What's certain is that ________ is the future of job recruiting and development, and understanding how to make sense of it will be critical to a company's success.
2. Recruiters can now gather a wealth of ________ from job boards. Objective information can be collected from ________ to help recruiters use this popular recruitment medium optimally, and craft the best recruitment messaging possible.
3. Recruiters are able to target specific ________ through data, to reach the highest likelihood of enlarging the ________ for each position.
4. Recruiters have to stop looking to the past for answers, and make the big switch to — ________. Predictive analytics are found in the ________ that you already have; they are simply the ________ of those data points. Predictive analytics help recruiters ask and answer the right questions.

Content Questions

1. What competencies should a recruiter have?
2. What are the characteristics of the recruitment space?
3. What happens if a company is experiencing a high turnover?
4. How to solve the problem caused by high turnover?

Answers

Blank Filling

1. big data
2. actionable data; job postings
3. talent sources; talent pool
4. predictive analytics; recruiting data; connection

Content Questions

1. Recruiters will not have to be analytical, mathematical clairvoyants, but they will need to be armed with the appropriate reporting and analytics tools.
2. The recruitment space is fast moving and heavily reliant on technology.
3. It's likely that there will be an entrenched, reoccurring problem that's causing the issue.
4. Data can help find solutions to these problems and it's because this that analytics within HR has a high adoption level.

参考译文

人力资源专员和招聘人员的工作正在发生着巨大的变化。就连一些轻松的机械化的工作也都在随着数据驱动的变化而不断地发展。当招聘人员掌握并学会一种工具时，便会出现另一种复杂的工具来替代它。当下的招聘工具是大数据。大数据几乎可以应用到所有领域，那么现在我们就来讨论一下大数据以及它在招聘营销中的应用。

“可以肯定的是，未来开展招聘工作一定会使用大数据，这一点对于公司能否成功是至关重要的。如今，大数据正在帮助快速发展的公司找到适合他们的工程师、开发人员和高管。”里维埃拉公司的 Michael A. Morell 说。

随着招聘工作的开展，招聘人员正在学习传统营销人员所需要的几项技能，他们也同时担任着数据分析师（不用担心，分析工具会帮助解决大部分工作）。招聘人员现在正在使用大数据下的招聘营销来战略性地吸引和留住优质人才——这引起了很多企业领导的关注。

招聘不再像“蒙着眼睛给驴贴尾巴”游戏

张贴公告然后祈祷结果的时代已经结束，招聘人员现在可以从招聘板块上收集大量可操作的数据。可以从招聘信息中收集客观信息，来帮助招聘人员熟练地使用这种热门招聘媒介，并制作出最佳的招聘信息。然后，找到最合理高效的将就业安排和展示信息结合起来的方式。

接下来，招聘人员可以通过得到的数据来定位具体的人才来源，尽可能地扩大每个职位的人才库。以前，对人才来源的寻找是有风险的，无目的地寻找会增加支出，可能意味着在没有投资回报率的情况下消耗预算，这就让你很被动。

现在，在招聘过程中应用大数据会帮助招聘人员创建自由流动的人才管道，并提高效率。职位公告板不再是令人沮丧又昂贵的“蒙着眼睛给驴贴尾巴”的游戏。招聘人员现在还可以使用智能化的作业平台。这一切都是为了更加有效地分配公司的招聘预算。

招聘中的魔术 8 球

到最近为止，招聘人员一直沉迷于招聘指标的一小部分。现在招聘人员不得不停下来改变过去的想法来进行预测分析。在已经拥有的招聘数据中就可以看到预测分析，它们只是这些数据点的连接，可以帮助招聘人员提出并解答问题。

- 在洛杉矶聘请 IT 专业人员需要几天的时间？
- IT 专业人士在哪一天，什么时候申请工作?
- 来自洛杉矶的 IT 专业人士最常应聘哪些职位？
- 来自洛杉矶的 IT 专业人士，在传统互联网（不是移动互联网）网站上应聘的放弃率是多少？

大数据可以在公司需要时帮助招聘人员瞄准、吸引和签署公司需要的人才。大数据连接着给招聘人员提供答案的数据点。大数据能够揭示我们在表面上看不到的趋势。这种知识使招聘人员具有自信的敏捷性，从而使招聘人员更准确地调整自己的方法和预算，从而实现最佳状态下的招聘营销工作。

大数据发展已被证明是有效的，并将进一步延续下去。德勤公司最近的研究显示，

57%的人力资源部门增加了他们对分析的支出。这是采用分析工具，将以前无用的数据变为可操作的和客观的信息的直接结果，这会让现在招聘人员拥有准确的数据来推翻以前的所有假设。

如前所述，招聘人员不再是分析性数学的学习者，而是分析工具的使用者。分析工具可以将数据转化为有用的信息，从而实现这种高度针对性的招聘营销。

凯利·罗宾逊

凯利·罗宾逊是 Broadbean 科技公司的创始人和首席执行官，Broadbean 科技公司是一家基于数据分析的招聘公司。Broadbean 科技公司在美国、欧洲和澳大利亚设有办事处，在全球范围内有很大的影响力。公司始终保持初心，坚持“轻松愉快工作”的理念，凯利注重领导力和企业文化，并尽力减少员工之间的摩擦。

大数据在招聘中的重要性

招聘领域发展迅速，并且越来越依赖技术。它率先应用在社交媒体领域，现在是在大数据和数据分析领域。

大数据已被用于各种行业，帮助做出更好的决策，寻找增加收入和推动增长的新途径。它虽然不是公司招聘人才所需的唯一工具，但它无疑为公司提供了另一种选择，可以设计和实施新的招聘战略。

如果一家公司的离职率很高，说明这家公司很可能会存在某种问题，可能是通勤时间、难应付的上司，也可能是工作时间不灵活，数据可以帮助找到问题的根源，正因为如此，分析技术在人力资源部门被广泛使用。

大数据在公司人力资源管理方面取得了长足进步，可以帮助预测岗位需求、识别优秀人才，预测分析哪里会得到求职者的积极的回应，发现求职者作出回应的具体时间，利用巨大的数据池发现有价值的趋势，设计出吸引最佳人才的职位描述。

Gild 这样的初创公司也吸引了许多人的目光，在招聘网页开发人员时，通过检查他们设计的代码和专业知识来帮助招聘人员深入了解应聘者，让招聘人员对其工作能力有一个准确的了解。尽管现在集中在一个领域，但在不久的将来，它的应用将更为广泛。

如前所述，技术正在人力资源领域产生重要的影响。诸如 Resumator 这样的软件，是对现有的人力资源体系的有效补充，技术正在对人力资源领域进行重塑，使公司能够吸引到最好的人才。

Text B

Big Data — the collection of larger than average datasets that require unconventional storage, processing, and analysis methods, has revolutionized nearly every field of business, from marketing to manufacturing. Big Data can provide those firms that develop the infrastructure to analyze and act on the patterns and insights contained in these datasets, with a source of competitive advantage in any industry. This infrastructure includes the technology to

New Words and Expressions

revolutionize/ˌrevəˈluːʃənaɪz/ *v.* 使发生革命性剧变

aggregate, process, and analyze various datasets, and the personnel to perform these operations, which marketing research firm Gartner estimates will be a $232 billion dollar industry by 2016. As more and more firms invest in Big Data infrastructure and integrate it into their existing internal operations, such personnel are in high demand these days. Firms often find them with the help of Big Data-driven recruiting procedures. Indeed, Big Data has transformed the world of recruiting; and it may help you find the talent you need, in each area of your business.

Big Data, or people analytics, as it is known when applied to recruiting, provides recruiters with more data to analyze. Social media networks have become the first stop for many recruiters after receipt of a resume. However, people analytics encompasses more than just social media data mining. Indeed, it encompasses even more than just back-end software or personnel. People analytics is also an orientation — an attempt to create a complete picture of a candidate long before they step foot in an office for an interview. An applicant's entire online presence, their use of a firm's recruiting database, their customer or non-customer status, their political affiliations, their smoking preferences, and other characteristics can all taken into consideration in this era of Big Data.

Benefits of Recruiting Using Big Data

The people analytics approach has tremendous advantages for recruiters. The proliferation of available information about candidates has made it possible for recruiters and human resources professionals to match an employee's professional and personal fit with their firm more closely to the firm's opening and corporate culture respectively. People analytics' tools and techniques allow firms to develop a much more complete profile of a candidate — far beyond a one-page cover letter and accompanying resume.

People analytics allows firms to move away from hiring based on subjective factors that may have very little to do with an employee's chances of success at that particular firm. The Big Data approach involves first determining what existing factors lead to employee success and retention, and hiring candidates who fall within those parameters. This approach makes it easier for recruiters and managers to justify new hires as well. And it works.

New Words and Expressions

aggregate/ˈæɡrɪɡət/ *v.*
使聚集

encompass/ɪnˈkʌmpəs/ *v.*
包含；包括

affiliation/əˌfɪliˈeɪʃən/ *n.*
联系

preference/ˈprefərəns/ *n.*
偏爱；爱好；喜爱

proliferation/prəlɪfərˈeɪʃən/ *n.*
扩散

fall within
应列入…范围内

Xerox recently used algorithm-driven recruiting techniques to reduce the attrition in its call centers by 20%.

Further, analyses of one's internal HR database, its strategic sales plan, and its accounts receivable, can yield insights about where a firm needs to hire to stay on top of existing orders. This insight allows firms to recruit proactively, rather than when they face a talent shortfall. Hiring proactively allows firms to spend the time necessary to select the right candidate, and avoid paying a premium for talent in moments of extreme organizational need. It also allows firms to develop strategic recruitment plans that incorporate a firm's broader hiring goals, such as building a diverse workforce.

People analytics can reduce your cost per hire, and your average time needed to fill open positions by making the recruiting process more efficient. Lastly, hiring using people analytics can align your compensation packages more closely with real market averages, by conducting analyses of publicly available salary information.

Recruiting Using Big Data

Big Data has given rise to a number of recruiting techniques designed to make recruiting efforts more precise and accurate. While these techniques predate the rise of Big Data, the explosion of available information has led to the development of algorithm-driven recruiting software solutions (as well as firms that specialize in algorithm-driven recruiting); and helped refine the tools and techniques used specifically for recruiting. These tools and techniques include data mining, keyword filtering, and testing.

- Data mining

Data mining is a technique used by firms to aggregate data for a variety of different business purposes, including recruiting. Data mining can be used to analyze the internal data created by high-performing and/or longstanding candidates to search for insights into their performance and/or longevity. Data-driven firms like IBM, along with standalone data analysis firms like the California-based Cataphora, specialize in such statistical analyses, which can be used for internal recruiting and/or retention. By analyzing from where successful candidates have been hired can simplify the recruiting process as well. For example, a firm whose internal analyses have revealed that 49% of their top performers had

New Words and Expression

attrition/əˈtrɪʃən/ *n.*
消耗，损耗

shortfall/ˈʃɔːtfɔːl/ *n.*
不足之数；缺口；差额

keyword filtering
关键词过滤

longevity/lɒnˈdʒevəti/ *n.*
长期供职

standalone/ˈstændəˌləʊn/ *adj*
单独的，独立的

their initial contact with a recruiter from Viadeo, may lead the firm to reduce advertising on LinkedIn, and instead ramp up recruitment efforts on the French social networking site.

Recruiters and human resources professional can also combine data mining with predictive analytics — the use of statistical methods and techniques to forecast the probability of a likelihood occurrence using historical data, to generate predictions about a candidate's likely tenure with the firm should they be hired. These insights can also be used to provide parameters for the recruiting of external candidates.

Data mining, or as some recruiters call it "talent mining" can be done manually or automatically online. Individual recruiters and/or software can search online resume databases (internal or external), professional social network profiles, or other websites of interest for personnel who might be a match for an opening.

Social networks, in particular, capture significant information about an individual. Recruiters can determine not only whether a candidate might be a good fit for the culture of the firm, but also whether they might be successful there, by assessing this information against internal profiles of high performing candidates. For example, a firm's highest performers may spend a small amount of time on a single social network. A candidate who spends considerable time on multiple social networks might raise some flags. Alternatively, a social network might indicate that the candidate is engaged in activities that might impair their productivity, such as excessive drinking or high-risk hobbies, such as extreme sports. These insights can be helpful to the diligent recruiter.

- Keyword filtering

Using desired skills and other characteristics as keywords, recruiters can run searches in popular search engines, on professional and non-professional search engines, in public or private online communities, and on other online properties. This can yield promising leads, who recruiters can contact for an informational or formal interview.

Keyword filtering is also helpful when screening out applicants who have applied for a position through a web-based talent management application (either proprietary or from a third-party recruiter). Recruiting software automatically scans

New Words and Expressions

parameter/pəˈræmɪtə(r)/ *n.* 参数；限制因素；决定因素

submitted resumes and cover letters for specific keywords, rejecting those without them, and returning to recruiters only the candidates who fit the job description on paper.

- Testing

More and more, testing is used in the hiring process. Usually, pre-screened applicants are invited to take a skills test, a personality test, or both. Skills tests are used to authenticate the skills listed in one's job application, but also can be used to test those not listed, such as soft skills. Personality tests are used to assess a candidate's fit with the firm's culture, as well as soft skills. Personality tests have been around for a long time, but the combination of computer-assisted testing, and data-driven approached to psychology, make these tests much more sophisticated and precise.

Increasingly, both skills and personality tests are assessed against internal analyses of high performing employees. For example, an advertising firm may find success with candidates who work well in a team and possess a high degree of digital fluency, regardless of the job opening. They may in turn offer measure all candidates for an opening against skills and personality tests they mandate during the hiring process.

It is not uncommon for candidates for senior positions in all industries (and even some junior level positions in industries such as finance) to be given one or multiple, skills tests, and a personality test, during multiple interview rounds. These tests provide hiring managers with more data points, alongside the job application, the interview(s), online data, and other publicly available information, against which to measure candidates.

New Words and Expressions

precise/prɪˈsaɪs/ *adj*.
精确的

Note:

The text is adapted from the website:
https://www.cleverism.com/best-uses-big-data-recruiting/.

参考译文

大数据——大规模数据的收集需要非常规的存储、处理和分析方法，几乎从市场营销到制造业的各个领域都发生了革命。大数据可以为那些开发基础设施的公司提供分析。大数据分析的结果可以指导公司决策，同时也是任何行业竞争优势的来源。该基础设施包括聚合、处理和分析各种数据集的技术，以及执行这些操作的人员，市场研究公司 Gartner 估计到 2016 年市场规模将达到 2320 亿美元。随着越来越多的企业投资大数据基础设施，并将其整合到现有的内部操作中，大数据从业者的需求日益高涨。公司经常在大数据驱动招聘程序的帮助下找到他们。事实上，大数据改变了招聘的世界；并且可以帮助你在每个业务领域找到所需的人才。

应用于招聘时，大数据或人员分析为招聘人员提供了更多的数据进行分析。收到简历后，社交媒体网络已经成为许多招聘人员的第一站。然而，人员分析不仅仅涉及社交媒体的数据挖掘，事实上，它不仅仅包括后端软件或人员。人员分析也是一个发展方向——在他们踏入办公室进行面试之前，招聘人员会试图创建候选人的完整画像。申请人的整个在线状态，他们对公司的招聘数据库的使用、他们处于客户或非客户状态、他们的政治立场、吸烟偏好和其他特征，都是在这个大数据时代的考虑因素。

使用大数据招聘的好处

人员分析方法对招聘人员具有巨大的优势。有关候选人的现有信息的激增使得招聘人员和人力资源专业人员尽可能将员工的专业和个人素质与公司的开放和企业文化匹配起来。人员分析工具和技术允许公司开发更完整的候选人资料——远远超出了一页的求职信和附带的简历。

人员分析使企业在招聘时摆脱主观因素，这些因素可能与员工在该公司取得成功的机会无关。大数据方法首先确定现有员工成功的因素参数，并招聘符合这些要求的候选人。这种方法使招聘人员和管理人员更容易找到合适的员工。这的确奏效了。施乐公司最近采用了算法驱动的招聘技术，将呼叫中心的员工人数减少 20%。

此外，对内部人力资源数据库、战略销售计划及其应收账款的分析可以让我们了解企业需要聘用哪些人，以保持现有订单的最高水平。这种洞察力使企业能够主动招聘，而避免陷入人才短缺。招聘的主动性允许公司有时间选择正确的候选人，并避免在公司急需人才时支付额外的费用。它还允许公司制定战略招聘计划，纳入公司更广泛的招聘目标，例如，建立多样化的员工队伍。

人员分析可以降低招聘成本，通过更有效率的招聘流程节省招聘时间。最后，通过对公开的薪资信息进行分析，雇用人员分析可以使薪酬方案与实际市场平均水平更吻合。

使用大数据进行招聘

大数据引发了一些招聘技巧，旨在使招聘工作更加严格和精确。虽然这些技术早于大数据的出现，但大量的可用信息带来了算法驱动的招聘软件解决方案（以及专门从事算法驱动型招聘的公司）的发展；并帮助改进了专门用于招聘的工具和技术。这些工具和技术包括数据挖掘、关键字过滤和测试。

- 数据挖掘

数据挖掘是企业用于为各种不同业务目的（包括招聘）汇总数据的技术。数据挖掘可用于分析由高绩效和/或长期候选人创建的内部数据，以考察他们的长期工作表现。数据驱动的公司（如 IBM），以及独立的数据分析公司（如加州的 Cataphora），专门从事这种统计分析，可用于内部招聘。通过分析候选人获得成功的方式，可以简化招聘过程。例如，一家公司的内部分析显示，49%的最佳表现者都是通过与 Viadeo 的招聘人员签订第一份合同招聘入职，这可能会导致该公司减少 LinkedIn 上的广告业务，取而代之的是在法国社交网站上加大招聘力度。

招聘人员和人力资源专业人员还可以将数据挖掘与预测分析结合使用——在研究历

史数据后，采用统计方法和技术对可能性发生概率进行预测，以此公司可以对员工的未来合同期进行推测，并决定是否继续雇用他们。同时也可以用于对外部人员的招聘。

数据挖掘，或者一些招聘人员称之为“人才挖掘”，可以手动或自动在线完成。个人招聘人员和/或软件可以搜索在线简历数据库（内部或外部）、专业社交网络配置文件或与空缺职位相匹配的人可能感兴趣的网站。

特别是社会网络可以抓取有关个人的重要信息。招聘人员不仅可以确定候选人是否适合企业的文化，而且还可以通过高绩效候选人的内部资料评估他们是否可能在那里取得成功。例如，一家公司的表现最好的人可能会花费少量的时间在一个社交网络上。在多个社交网络上花费相当长时间的候选人可能会更被留意。或者，社交网络可能表明候选人从事可能损害其生产力的活动（如过度饮酒）或高风险的爱好（如极限运动）。这些都可以为勤奋的招聘人员提供参考。

- 关键字过滤

使用期望的技能和其他特征作为关键字，招聘人员可以在流行的搜索引擎、专业和非专业搜索引擎、公共或私人在线社区以及其他在线资源中进行搜索。这可以找到有希望胜任的人员，招聘人员可以与其联系进行正式面试或非正式面谈。

通过基于网络的人才管理应用程序（专有人员或第三方招聘人员）筛选申请职位的申请人，关键字过滤也是有帮助的。招聘软件会自动扫描提交的简历和封面信件以获得特定的关键字，拒绝没有这些材料的人员，并且仅向符合工作描述的候选人提供面试机会。

- 测试

招聘采用越来越多样的测试。通常，预先筛选的申请人被邀请参加技能测试、个性测试或两者都参加。技能测试用于认证工作申请中列出的技能，也可用于测试未列出的技能，如软技能。个性测试用于评估候选人与公司文化的合适性以及软技能。人格测试已经存在了很长时间，但计算机辅助测试与数据驱动相结合的心理测试使得这些测试更为复杂和精确。

越来越多地针对高绩效员工的内部分析对技能和人格测试进行评估。例如，不管职位是否空缺一家广告公司都可能会找到成功的候选人，他们在团队中工作良好，拥有高度的数字素养。反过来，他们可以向所有候选人提供入职培训，在招聘过程中衡量他们所要求的技能和个性测试。

所有行业的高级职位（甚至金融行业的初级职位）的候选人都会获得一次或多次技能考试和个性考试，这是常见的现象。除了工作申请、面试、在线数据和其他公开信息，这些额外的测试为招聘经理提供了更多的数据，用于评价候选人。

Chapter 8

Big Data in Gaming Industries

Text A

Big Data in Gaming Industry Improves Gaming Experience

EA games has more than two billion video game players in the world, who generate approximately 50 terabytes of data each day. The gaming industry does $ 20 billion in annual revenue in America alone of which 2 billion in sub-category social games. In the USA, the gaming industry is bigger than the movie industry (with an annual amount of $ 8 billion spent on movie tickets). The world of gaming is big, growing rapidly and taking full advantage of the big data technologies. Gaming companies can drive customer engagement, make more money on advertising and optimize the gaming experience among others with utilizing the big data in gaming industry.

An Improved Customer Experience

As with any organization, also the 360-degrees customer view is important for the gaming industry. Fortunately, gamers leave a massive data trail when they play a game. Whether it is an online social game connected via Facebook, a game played on an offline PlayStation or a multi-player game via the Xbox, a lot of data is created in different formats when gamers start playing. They create massive data streams about everything they do within a game. How they interact, how long they play, when they play, with whom, how much they spend on virtual products, with whom they chat etc. If the gaming profile is linked to social networks or a gamer is asked

New Words and Expressions

terabyte/ˈterəbaɪt/ *n.* 太字节（TB）

profile/ˈprəʊfaɪl/ *n.* 简介

to enter demographical data the information can be enriched with what the gamer likes in real life and gaming companies can adapt the game in real life to the profile of the gamer.

Based on all that data targeted in-game products can be offered that have a high conversion rate. Just like on e-commerce websites were products are recommended based upon what other customers bought, this can also be done within the gaming environment. Recommending certain features that other players also bought that can be bought with a product or recommending certain virtual products based on the level the gamer is in. This can result in an increased up-sell or cross-sell ratio and additional revenue.

Engagement can also be increased if analytics show that a player will abandon the game if the first levels are too difficult or if later levels are too easy. Data can be used to find bottlenecks within the game, where many players fail the tasks at hand. Or it can be used to find the areas that are too easy and need to be improved. Analyzing millions of player data gives insight into which elements of the game are most popular. It can show what elements are unpopular and requires action to improve the game. Constant engagement is vital and with the right tools the right reward can be provided at the right moment for the right person within the game to keep a player engaged.

Big data in gaming industry technologies also help to optimize in-game performance and end-user experience. When for example the databases and servers of the games have to cope with a steep increase in online players, it is important to have sufficient capacity. With big data it is possible to predict the peaks in demand to anticipate on the required capacity and scale accordingly. This will improve the gaming experience (who likes a slow game) and thus the end-user experience.

To Deliver a Tailored Gaming Experience

Games that are developed for different consoles or devices (tablets vs smartphone or Xbox vs PlayStation) can result in a different playing experience. When all data is analyzed, it can provide insights in how players play the game on different devices and whether there is a difference to be solved.

Big data also enable to show tailored individual in-game advertising corresponding with the needs and wished of the player.

New Words and Expressions

e-commerce/ˌiːˈkɒmɜːs/ *n.*
电子商务

console/kənˈsəʊl/ *n.*
操纵台

cross-sell
交叉销售

tailored/ˈteɪləd/ *adj.*
特制的，特供的

With all the big data in gaming industry, created by gamers, a 360-degree in-game profile can be created that, when combined with open and social data from the gamer, can give insights in the likes and dislikes of that gamer. This information can be used to show only those advertising within the game that matches the profile of the gamer resulting in a higher stickiness factor of the advertising and more value for the advertiser and subsequently more revenue for the game developer.

There are ample opportunities for game developers to improve the gaming experience with big data, drive more revenue and improve the game faster and better. Game developers should therefore not miss out on big data, because the benefits for the developer as well as the player are too big to ignore when looking at total big data in gaming industry.

New Words and Expressions

ample/ˈæmpəl/ *adj.*
足够的，充足的

Note:

The text is adapted from the website:

http://playbook.amanet.org/big-data-in-gaming-industry- improves-gaming-experience/.

Comprehension

Blank Filling

1. The world of gaming is big, growing rapidly and taking full advantage of the big data technologies. Gaming companies can drive customer ________, make more money on __________ and optimize the gaming _________ among others with utilizing the big data in gaming industry.
2. A lot of data is created in different formats when gamers start playing. They create massive ___________ about everything they do within a game.
3. If the gaming profile is linked to social ________ or a gamer is asked to enter _________ data the information can be enriched with what the gamer likes in ______ life and gaming companies can adapt the game in real life to the profile of the gamer.
4. Analyzing millions of __________ gives insight into which elements of the game are most popular. It can show what elements are unpopular and requires action to ______ the game.
5. Big data in gaming industry technologies also help to _______ in-game ________ and end-user __________.
6. When all _____ is analyzed, it can provide insights in how players play the game on different _____ and whether there is a difference to be solved.

7. With all the big data in gaming industry, created by gamers a 360-degree _______ can be created that, when combined with ___________ data from the gamer, can give insights in the ____________ of that gamer.
8. Game developers should therefore not miss out on _________, because the _______ for the developer as well as the player are too big to _______ when looking at total big data in gaming industry.

Content Questions

1. What does the game companies use the big data to do?
2. What kind of players' information can the game company get?
3. What is the effect of big data in the game industry?
4. How does the game company get the player's favorite? Why?
5. What is necessary to ensure that players participate in?
6. What is data for in gaming industry?

Answers

Blank Filling

1. engagement; advertising; experience
2. data streams
3. networks; demographical; real
4. player data; improve
5. optimize; performance; experience
6. data; devices
7. in-game profile; open and social; likes and dislikes
8. big data; benefits; ignore

Content Questions

1. Gaming companies can drive customer engagement, make more money on advertising and optimize the gaming experience among others with utilizing the big data in gaming industry.
2. How they interact, how long they play, when they play, with whom, how much they spend on virtual products, with whom they chat etc.
3. An Improved Customer Experience, To Deliver a Tailored Gaming Experience.
4. Analyzing millions of player data gives insight into which elements of the game are most popular. Because it can show what elements are unpopular and requires action to improve the game.
5. Constant engagement is vital and with the right tools the right reward can be provided at the right moment for the right person within the game to keep a player engaged.

6. It can provide insights in how players play the game on different devices and whether there is a difference to be solved.

参考译文

游戏行业的大数据改善了游戏体验

艺电公司（EA）的游戏在全球有超过 20 亿的视频游戏玩家，这些玩家每天生成大约 50 太字节（TB）的数据。游戏行业在美国的年收入为 200 亿美元，其子类别社交游戏年收入达到 200 亿美元。美国的游戏行业比电影行业繁荣（美国人每年花费约 80 亿美元买电影票）。游戏世界规模庞大、发展迅速，同时充分利用大数据技术。游戏公司通过利用游戏行业的大数据推动客户参与，在广告上赚更多的钱并优化游戏体验。

改善客户体验

与大多数组织一样，360 度的客户视角对于游戏行业也是非常重要的。幸运的是，玩家进行游戏时会留下大量的数据痕迹。无论是通过 Facebook 连接的在线社交游戏，还是在离线的 PlayStation 上玩的游戏或使用 Xbox 的多玩家游戏，玩家开始游戏时都会以不同的格式创建大量数。游戏玩家们制造了关于他们在游戏中所做一切的大量数据流。他们的交互方式、玩了多长时间、什么时间玩的、和谁玩、在虚拟产品上花费了多少钱以及他们与谁聊天等。如果游戏资料与社交网络相关联或者要求玩家输入统计资料，信息因为玩家在现实生活中喜欢的内容得到丰富，并且游戏公司可以根据玩家现实生活情况编写游戏。

基于所有这些数据，游戏可以有针对性地提供高转换率的产品。就像在电子商务网站上一样，商品是根据其他顾客购买的产品推荐的，这也可以在游戏环境中完成。推荐其他玩家也购买的某些功能，或根据玩家所在的级别推荐某些虚拟产品，这可能会使产品的畅销率或交叉销售率以及额外收入得以增长。

如果一个游戏角色的初始等级太难，或角色在高等级时游戏内容太简单，就会使玩家放弃游戏，按照此分析进行调整，就可以增加玩家的参与度。可以使用数据来查找游戏中的瓶颈，例如许多玩家无法完成的任务，或者玩家很容易找到游戏中某个区域，对这些情况就需要改进。分析数百万玩家的数据可以深入了解游戏的哪些元素最受欢迎，也可以显示什么元素不受欢迎，需要采取行动来改善游戏。持续地参与至关重要，通过正确的工具，可以在适当的时机为游戏中合适的人员提供合理的奖励来保证玩家的参与。

游戏行业中的大数据技术也有助于优化游戏内部的性能和最终用户体验。例如，如果游戏的数据库和服务器必须应付在线玩家的急剧增长，则它们必须具有足够的存储空间。通过大数据，可以预测需求峰值，从而预期所需的存储空间和性能规模。这将改善游戏体验（没人会喜欢慢游戏），从而改善最终用户的体验。

提供量身定制的游戏体验

为不同的游戏机或设备开发的游戏（平板电脑、智能手机、Xbox 和 PlayStation）可以产生不同的游戏体验。采用大数据平台时，可以得到玩家们在不同设备上玩游戏的分析结果，以及是否有待解决的差异。

大数据还能够显示符合玩家需求和愿望的量身定制的个人游戏内置广告。通过游戏玩家在游戏行业中创建的所有大数据可以创建出一套360度的游戏内容资料——当与游戏玩家的开放的社交数据相结合时，可以深入了解玩家的喜好。信息可用于只显示与游戏者的配置文件相匹配的广告内容，使得广告的黏度更高，给广告商更高的价值，随后游戏开发商的收入也将增加。

游戏开发人员有充足的机会通过大数据改善游戏体验、带来更多的收入并提高游戏的速度和质量。纵观大数据在游戏业的使用，开发者以及玩家都从中获益。因此，对游戏开发人员而言，大数据不容错过。

Text B

Back in 2013, the Staples Center sold out in under an hour, setting what must be some unofficial record for ticket sales. However, people weren't buying tickets to see the Lakers or the Kings play. The center was actually hosting the League of Legends Championship. For those of you who don't know, League of Legends is a popular, free-to-play, fantasy combat game. So yes, you heard correctly. Over 10,000 paid to watch people play video games.

The electronic gaming industry has come a long way since the days of Pong and Pac-Man. It's now valued at more than 90 billion dollars. And it isn't just the major companies like Sony, Microsoft or Electronic Arts (EA) who are contributing to the industry. There are thousands of smaller developers and new designers creating games for newer platforms, like social media sites or mobile devices.

Because there's so much money on the line, and a saturated market of players and developers, there's an extreme amount of competition. Gamers only have so much time, and so much money, meaning developers are constantly fighting to get their games into the hands of these players. Every minute and dollar spend with one developer takes the same away from someone else. Not to mention, gamers expect more than ever before. Graphics and gameplay are reaching new levels. The limits are constantly being pushed, and gamers won't settle for mediocre. Many big games pushed out on major systems cost a small fortune, meaning if they fail, the company behind it may be toast. And now, with the prevalence of social media and online reviews, all it takes is a few influential

New Words and Expressions

saturated/ˈsætʃəreɪtɪd/ *adj*.
饱和的

prevalence/ˈprevələns/ *n*.
流行

mediocre/ˌmiːdiˈəʊkə(r)/ *adj*
普通的；中等的

gamers to give a poor rating, and a game can fall flat within only weeks of its release.

So obviously, there's a lot on the line. Developers don't want to be gambling hundreds of millions of dollars on a whim. That's exactly why more and more companies are relying on gamer data to increase the chances of developing a popular game that'll sell. In this sense, gaming companies may want to take a page out of Netflix's book. Netflix collects massive amounts of information on its customers and viewing habits. That way, when it releases a series, it's already pretty confident already that it's going to do well, like with "House of Cards." Almost all of today's entertainment consoles, or even mobile gaming platforms, come standard with internet connectivity. This isn't just to allow gamers to play against other gamers from all over the world. Online gaming certainly has a dual-purpose. Creating online accounts allows developers to learn the types of games people are playing, and how they're being played. With this information they can learn the trends and customize games or gameplay to fit with demands, and vastly improve the chances of creating a successful game.

However, big data isn't just booming for the big gaming companies. It isn't also just for creating the right kind of games either. One of the biggest challenges aside from making a games, is learning how to monetize them. This is especially true for games on social media or mobile devices, like smartphones. There are many different ways to monetize games, like merchandising or offering pay-to-play, and it's important to determine the right method for your audience.

Developers looking to stay on top, and remain aware of gamer demands and trends, will need to invest in the right tools. Analyzing millions of users and hundreds of metrics across multiple titles and platforms is obviously problematic. Most large companies, especially within the gaming industry, aren't looking to have big data gaming analysis chew up all of their time. They're in the business of developing games and gaming platforms. Going with the right service will ensure your organization's needs are being met, without taking too much time away from core business functions.

New Words and Expressions

whim/wɪm/ *n.*
突然的念头，冲动

connectivity/ˌkɒnekˈtɪvəti/ *n.*
连接性能

Note:

The text is adapted from the website:

http://dataconomy.com/2015/02/big-data-takes-over-the-gaming-industry/.

参考译文

2013年，斯台普斯球馆在一个小时内卖完了所有的票，创造了最快售票纪录。不过，人们并不是买门票来看湖人队或是国王队的比赛。实际上该中心举办的是《英雄联盟》锦标赛。《英雄联盟》是一个广受欢迎的、免费玩的、虚幻的战斗游戏。所以，你没有听错，超过10 000人买票观看视频游戏比赛。

自从有了诸如乒乓球和吃豆人等电子游戏，游戏产业已经走了很长的路，现在价值超过900亿美元。不仅仅是索尼、微软或艺电（EA）等大型企业对行业做出了贡献，成千上万的小型开发人员和新设计师也都为较新的平台（如社交媒体网站或移动设备）在编写游戏。

因为线上有很多利益，市场上玩家和开发商饱和，所以竞争非常激烈。玩家只有有限的时间和钱，这意味着开发人员得不断地争取把他们的游戏宣传到玩家的手中。开发商的每一分钟和每一块钱都与其他人一样；更不用说，玩家比以往任何时候的期望都多。界面和游戏玩法正在达到新的水平。限制不断被推动，玩家不会将就。许多在大型系统上推出的大型游戏都花费巨大，这意味着如果它们失败了，那么背后的公司可能就完了。而现在，随着社交媒体和在线评论的普及，一些有影响力的玩家会给予不好的评价，最终导致游戏在发布后的几个星期内就彻底失败。

很显然，在线的人很多。开发商不想在一时冲动下赌上数亿美元。这就是为什么越来越多的公司依靠玩家数据来增加开发流行游戏的机会。在这个意义上，游戏公司可能想从Netflix公司①的预定记录中获得一些经验。Netflix收集大量关于客户的信息和观看习惯。因此，当Netflix发行一系列视频产品之前，它就已经信心十足，就像发行电视剧House of Cards（纸牌屋）取得成功一样。现在几乎所有的游戏机甚至是移动游戏平台都进行了标准化的互联网连接，这不仅仅是为了玩家能与来自世界各地的其他玩家玩游戏，开发一款在线游戏通常具有双重目的：一是创建在线账户，这使得开发人员了解人们正在玩的游戏类型；二是了解游戏的具体方式。借助这些信息，他们可以了解游戏趋势并根据需求定制游戏，这样大大提高了创造成功游戏的机会。

然而，大数据不仅仅促进了大型游戏公司的蓬勃发展。它也不是仅仅为了创造合适的游戏而存在的。除了制作游戏之外，最大的挑战之一是知道如何通过它们获利。这对于社交媒体或移动设备（如智能手机）的游戏尤其如此。通过游戏获利有很多不同的方法，如商品销售或提供付费游戏，重要的是为玩家探索正确的方法。

---

① Netflix公司成立于1997年，是一家在线影片租赁提供商，主要提供Netflix超大数量的DVD并免费递送，总部位于美国加利福尼亚州洛斯盖图。

开发人员想保持领先地位并意识到玩家的需求和趋势，需要投资于正确的工具，以此来分析数百万用户和数百个游戏平台，但显然从如此庞大的数据得到分析结果是非常困难的。大多数大型公司，特别是游戏行业中的公司，并不希望利用大数据分析游戏占用他们的时间，他们更希望能够专注于开发游戏和游戏平台，使用正确的服务将确保公司的需求得到满足，而不需要太多的时间远离核心业务功能。

Chapter 9

Big Data in Education

Text A

Colleges and universities are finally ditching legacy database systems and moving on to managing big data and its applications. This only goes to show how big data's adoption will be insuppressible across various industry sectors.

Big data is making bigger inroads into the education industry

Colleges and universities are not only inundated with data from legacy systems but have also begun to link disparate information from across the campus. The application of data-driven decision making has begun to permeate all aspects of campus life and operations, as enterprising leaders harness predictive analytics to tackle bottleneck courses, power advising initiatives and share best practices with their peers. We look at some features here that big data application might be able to provide in education sector.

Improved Student Results

The overall goal of Big Data within the educational system should be to improve student results. The answers to assignments and exams are the only measurements on the performance of students. During his or her student life, every student generates a unique data trail. This data trail can be analyzed in real-time to deliver an optimal learning environment for the student as well as to gain a better understanding in the individual behavior of the students.

It is possible to monitor every action of the students — how

New Words and Expressions

ditch/dɪtʃ/ *v.*
抛弃，遗弃，丢弃

insuppressible/ˌɪnsəˈpresəbl/ *adj.* 抑制不住的

inundate/ˈɪnʌndeɪt/ *v.*
淹没

assignment/əˈsaɪnmənt/ *n.*
任务，作业

trail/treɪl/ *v.*
追踪

long they take to answer a question, which sources they use, which questions they skipped, how much research was done, what the relation is to other questions answered, which tips work best for which student, etc. Answers to questions can be checked instantly and automatically (except for essays perhaps) to give instant feedback to students.

In addition, Big Data can help to create groups of students that prosper due to the selection of a group. Students often work in groups where they may not be complementary to each other. With algorithms, it would be possible to determine the strengths and weaknesses of each individual student based on the way a student learned online, how and which questions were answered, the social profile etc. This will create stronger groups that will allow students to have a steeper learning curve and deliver better group results.

Create mass customized programs

All the data will help to create a customized program for each individual student. Big Data allows for customization at colleges and universities, even if they have 10,000 students. This can be created with blended learning; a combination of online and offline learning. It will give students the opportunity to develop their own personalized program, following those classes that they are interested in, working at their own pace, while having the possibility for (offline) guidance by professors. Providing mass customization in education is a challenge, but algorithms make it possible to track and assess each individual student.

We already see this happening in the MOOC's (Massive Open Online Courses) that are being developed around the world. When Andrew Ng taught the Machine Learning class at Stanford University, generally 400 students participated. When it was developed as a MOOC at Coursera in 2011, it attracted 100,000 students. Normally this would take Andrew Ng 250 years to teach the same amount of students. 100,000 students participating in a class generates a lot of data that can deliver tremendous insights. Being able to cater for 100,000 students at once also requires the right tools to be able to process, store, analyze and visualize all data involved in the course. At the moment, these MOOC's are still mass made, but in the future they can be mass customized.

Reduce dropouts, increase results

When students are closely monitored, receive instant feedback

New Words and Expressions

complementary
/ˌkɒmplɪˈmentəri/ *adj.*
补充的；互补的

customization
/ˈkʌstəmaɪzeɪʃən/ *n.*
用户化，专用化，定制

dropout/ˈdrɒpaʊt/ *n.*
辍学者

and are coached based on their personal needs, it can help reduce dropout rates. Predictive analytics on all the data that is collected can give educational institutes insights in future student outcomes. These predictions can be used to change a program if it predicts bad results on a particular program or even run scenario analysis on a program before it is started. Universities and colleges will become more efficient in developing a program that will increase results, thereby minimizing trial-and-error.

Over the last decade, Georgia State coupled data analytics with college advising to eliminate the gap in graduation rates between low-income and minority students and the rest of its student body, while also raising their overall graduation rate by 22 points.

After graduation, students can still be monitored to see how they are doing in the job market. When the resultant insights are made public, it will help future students in their decision to choose the right university.

While big data is still in a very nascent phase, its advantages in every sector are being realized with every passing day. The Education sector will always continue to be one of the most important areas of development for any country. Incorporating big data methods in education is surely going to help the students and society by placing the right people at the right positions. It's our future, let's make it big.

New Words and Expression

eliminate/ɪˈlɪmɪneɪt/ *v.*
排除；消除

nascent/ˈneɪsənt/ *adj.*
新生的

Note:

The text is adapted from the website:

http://www.linkedin.com/pulse/big-data-making-bigger-inroads-education-industry-naveen-joshi.

Comprehension

Blank Filling

1. Colleges and universities are not only inundated with data from ________ but have also begun to link ________ information from across the campus.
2. During his or her student life, every student generates a unique ________. This data trail can be analyzed in real-time to deliver an optimal learning ________ for the student as well as to gain a better understanding in the individual ________ of the students.
3. With algorithms, it would be possible to determine the ________ of each individual student based on the way a student learned ________, how and which questions

were answered, the social _______ etc.

4. Providing mass customization in education is a challenge, but algorithms make it possible to _________ and _________ each individual student.
5. Predictive analytics on all the data that is collected can give educational institutes insights in future student _______. These predictions can be used to ______ a program if it predicts bad results on a particular program or even run __________ on a program before it is started.
6. After graduation, students can still be ________ to see how they are doing in the job market. When the resultant insights are made public, it will help future students in their ______ to choose the right university.

Content Questions

1. What are the advantages of applying large data in the education industry?
2. What is the overall goal of big data in the education system?
3. What is the data trail for during student's life?
4. How does big data help reduce dropout rates?

Answers

Blank Filling

1. legacy systems; disparate
2. data trail; environment; behavior
3. strengths and weaknesses; online; profile
4. track; assess
5. outcomes; change; scenario analysis
6. monitored; decision

Content Questions

1. Big data is making bigger inroads into the education industry.
   Improved Student Results.
   Create mass customized programs.
   Reduce dropouts, increase results.
2. The overall goal of Big Data within the educational system should be to improve student results.
3. The data trail can be analyzed in real-time to deliver an optimal learning environment for the student as well as to gain a better understanding in the individual behavior of the students.
4. When students are closely monitored, receive instant feedback and are coached based on their personal needs, it can help reduce dropout rates.

参考译文

高等院校终于脱离了传统的数据库系统，并开始采用大数据应用来管理数据，大数据在以下方面得以蓬勃地发展。

大数据正在逐步进入教育行业

高校不仅充斥着遗留系统的数据，还开始将来自整个校园的不同信息联系起来。数据驱动型决策的应用已经开始渗透到校园生活的各个方面，有进取心的领导者利用预测分析来处理瓶颈课程，咨询计划并与同行分享。让我们来看看大数据如何在教育领域得到应用。

提高学生成绩

教育系统中大数据的总体目标应该是提高学生成绩。作业和考试的答案是衡量学生表现的唯一标准。在学生生活中，每个人都会生成一个独特的跟踪数据。该跟踪数据可以实时分析，为学生提供最佳的学习环境，并更好地了解学生的个人行为。

它还可以监测学生们的每一个行动，例如，他们回答一个问题需要多长时间，使用哪些资料，跳过了哪个问题，进行了多少研究，这些问题与其他问题之间的关系是什么，哪些建议最适合哪个学生，等等。问题的答案可以立即自动检查（除了论文之外）并给学生反馈。

此外，大数据可以在学生群体中帮助他们筛选出一个完美的小组。学生就可以在彼此互补的群体中成组地去工作。通过算法，还可以根据学生在线学习的模式、回答问题的方式、社交概况等来确定每个学生的优缺点。这将创建更强大的团队，让学生拥有一个整体呈直线上升的学习曲线，并提供更好的团队成果。

创建大量自定义程序

海量数据将有助于为每个学生创建一个定制的计划。即使高校有10 000名学生，大数据允许高校为每个学生量身定制计划。这使创建混合学习即在线、离线学习组合的方式成为可能。它将让学生有开发自己的个性化课程的机会，按照他们感兴趣的课程和理想的速度投入学习，同时有可能离线获得教授的指导。在教育中提供大规模定制是一个挑战，但算法使得跟踪和评估每个学生成为可能。

目前，在全球范围内开发的MOOC（大规模开放在线课程）中实现了上述的定制计划。安德鲁·伍恩在斯坦福大学教授机器学习课时，一般有400多名学生参加。2011年MOOC在C语言课程上，吸引了10万名学生。同样情况下，安德鲁·伍恩需花费250年时间才能教同样数量的学生。参加课程的10万名学生产生了大量可以提供非凡见解的数据。能够同时照顾10万名学生，需要适当的工具才能处理、存储、分析和可视化课程中涉及的所有数据。目前，这些MOOC课程已经大规模展开，但未来仍可以大量定制。

减少辍学率，提升效果

当学生受到密切监测时，接收并即时反馈信息，根据个人情况进行辅导，可以帮助减少辍学率。收集的所有数据的预测性分析可以使教育机构洞察学生可能在未来取得的成就。

如果系统在场景分析中采用试错法预测出一些坏的结果，甚至在项目开始前对其运行方案分析，则可以使用这些预测来更改项目。高校将会开发一个高效且可以提高预测准确度的应用，从而最大限度地减少试错法带来的危害。

过去十年间，佐治亚州将数据分析与高校咨询建立起联系，旨在消除低收入和少数民族学生与其余学生之间毕业率的差距，同时将毕业率提高 22%。

毕业后，大数据系统仍然可以监测学生，看看他们在就业市场上的表现。当公开这些结论时，未来的学生可以依据这些结论来选择合适的大学。

虽然大数据仍处于初级阶段，但其在每个行业的优势正在逐渐实现。教育领域永远是所有国家最重要的发展领域之一。将大数据方法纳入教育，肯定会帮助学生和社会，把正确的人放在正确的位置。这是我们的未来，让我们一起实现。

Text B

School of Big Data: How Analytics Benefits Higher Education

For the past few years, big data has been making waves across nearly every industry. However, nowhere is this more true than the educational sector. Higher education institutions are typically some of the first adopters of new technology, and colleges and universities across the country have been shaping their educational, recruitment and retention programs thanks to insights gleaned from big data analytics.

What's more, with more employment opportunities for data scientists and analysts than ever before, many schools are offering new courses to ensure students are ready for their future careers.

Big data analytics has much to offer when it comes to higher education. Let's take a look at a few ways colleges, universities and other schools can leverage these processes to their full advantage.

Pulling from an array of internal sources

While nearly every organization in any industry likely has a wealth of informational assets at its disposal from which to mine data, this is particularly true for educational institutions. Schools obtain a treasure trove of information from current and prospective students — and this is only a single big data source. Educational organizations are also turning to older systems to gather details and analyze them for valuable insights, opening up new possibilities.

"Colleges and universities, inundated with data from legacy systems and incentivized by renewed accountability pressures, have begun to link disparate information from across the campus,"

New Words and Expressions

array/əˈreɪ/ *v.*
数组；队列，阵列；一大批

Bridget Burns, University Innovation Alliance's executive director, wrote for Forbes. "Historically limited to transactional data from registrars and student information systems, the application of data-driven decision-making has begun to permeate all aspects of campus life and operations — as enterprising leaders harness predictive analytics to tackle bottleneck courses, power advising initiatives and share best practices with their peers."

In this way, schools aren't just leveraging all of their available resources. Administrators are also seeking out innovative ways to apply analytics findings to processes all across the institution.

Setting sights on social media

Schools are looking beyond their own big data for further insights as well. Leveraging social media information has now become a more common trend. In this way, schools' recruitment teams and administrators can learn as much about a particular student or group of pupils as possible.

A recent Kaplan Test Prep survey found that 40 percent of admissions officers currently engage social media resources to get additional information about applicants. This process is also becoming more popular with scholarship funds as organizations seek to award monies to the most promising student candidates.

Geographical targeting: Hitting students where they live

Recruitment efforts have been especially impacted by big data, where analytics can help reveal where schools should focus their efforts and what kind of return on investment they can expect.

Similar to the retail industry, higher education institutions can reduce their marketing spending by creating more targeted campaigns that appeal to audiences in a specific area. For example, if big data insights show that students in certain cities not only apply regularly to a college, but are most often accepted, the organization can adjust its marketing efforts accordingly.

"If a university admissions office has a firm understanding of which geographical locations include the most applicants who enroll, it can cut marketing costs and produce enhanced results," Lauren Willison, Florida Polytechnic University director of admissions wrote in a guest post for IBM. "Rather than investing in

New Words and Expressions

permeate/ˈpɜ:mieɪt/ *v.*
渗透；弥漫

enterprising/ˈentəpraɪzɪŋ/ *adj.*
有开创能力的

seek to
设法

retail/ˈri:teɪl/ *n.*
零售，零卖

admissions office
招生办公室

enroll/ɪnˈrəʊl/ *v.*
（使）加入；注册

unfocused campaigns that target a wide audience, money can be invested in specific markets that are more likely to boost the university's yield rate."

Post-application: Selecting students to accept

Big data analytics doesn't end after a student has submitted their application. Fast Company contributor Neal Ungerleider reported that more institutions are also leveraging big data to help make decisions about which students will be accepted. Analysis of certain datasets can show which candidates are the most likely to succeed at the college or university, and which might be more prone to drop out or fail.

This type of predictive analytics is currently being used by Wichita State University, where it's helping administrators make better informed decisions. Research shows that the school's recruitment approach results in 96 percent accuracy in pinpointing which applicants are "high-yield" or will likely do well at the institution.

Identifying educational troubles

Analytics can also be used by schools to pinpoint which pupils might be struggling in their educational pursuits. One of the best ways this strategy has been applied is to identify troubles earlier in a student's academic career. For instance, if a student is underperforming in prerequisite classes, advisors can help guide them before they fall behind, fail a course or drop out of school.

"Instead of falling through the cracks, students receive an early intervention with solutions such as rearranging course loads or exploring other paths to a degree," Willison pointed out.

From big data to big dollars

With so many industries buzzing about the advantages that big data analytics can offer, more individuals are seeking to pursue careers in this field. As a result, institutions are putting more big data, data science and analytics courses and programs into place. Tech Republic recently published a list of the top 20 schools offering such education, with Carnegie Mellon, Stanford and Santa Clara University topping the list.

New Words and Expressions

yield rate
收益率

prone to
倾向于…

crack/kræk/ *n.*
裂缝

intervention/ˌɪntəˈvenʃən/ *n.*
介入，干涉

buzzing/ˈbʌzɪŋ/ *v.*
充满兴奋的谈话声

Requirements and considerations: An expert partner needed

Big data analytics can offer nearly endless opportunities to improve campus processes and enhance education. However, there are a few challenges and requirements that administrators should be aware of, not the least of which is data security.

Institutions must have a secure environment in which to gather and work with their data, particularly if information includes students' sensitive personal details. What's more, with so many likely disparate data sources, it's essential to consider the work involved in integrating and organizing these datasets.

New Words and Expressions

disparate/ˈdɪspərət/ *adj*.
完全不同的

essential to
对…必不可少

integrate/ˈɪntɪgreɪt/ *v*.
使一体化；使整合

Note:

The text is adapted from the website:

http://www.linkedin.com/pulse/school-big-data-how-analytics-benefits-higher-heather-short-davis.

参考译文

大数据下的学校：高等教育如何受益于数据分析

在过去的几年中，大数据在每一个行业持续不断地制造轰动。但是，还没有哪个地方比教育部门更深受其影响。高等教育机构通常是新技术的第一批采用者，得益于大数据分析的深入研究，全国各地的高等院校一直在制定其教育、招聘和留用计划。

此外，数据科学家和分析师的就业机会比以往任何时候都多，许多学校正在提供新课程，以确保学生为未来的职业做好准备。

大数据分析在高等教育方面有很大的贡献。下面介绍高等院校充分利用大数据的几种方式。

从内部源数组中得出重要信息

几乎任何行业的组织都有可能拥有丰富的信息资产，可从中挖掘数据，教育机构尤其如此。学校从现在和未来的学生中获得宝贵的信息，这是一个大数据源。教育机构也正在转变旧系统，收集细节并分析，寻找有价值的见解，开辟新的可能性。

大学革新联盟执行董事布里奇特 • 伯恩斯（Bridget Burns）为《福布斯》杂志写道："高等院校保存着大量历史系统数据，在不断革新和激励问责的压力下，已将这些来自整个校园的不同信息联系起来。"以前的数据仅限于注册和学生信息系统，如今数据驱动型决策的应用已经开始渗透到校园生活的各个方面——学校利用预测分析来处理瓶颈课程，提供咨询计划的动力并与同行分享最佳的实践方法。

这样，学校不仅仅是利用其所有的可用资源。管理员也在寻求创新的方法来将分析结论应用于整个机构的流程中。

着眼于社交媒体

学校正在进一步处理自己的大数据，以获得更进一步的洞察。目前，利用社交媒体信息已成为一个更为普遍的趋势。学校的招生团队和管理人员可以尽可能多地了解特定的学生或学生群体。

最近的 Kaplan 测试调查发现，40%的招生人员目前正在参考社交媒体的资源，以获得相关申请人的更多信息。这个过程也越来越受到奖学金基金的欢迎，因为组织想设法为最有希望的学生候选人提供奖金。

地理位置定位：了解目标学生的位置

招生工作受到大数据的影响，数据分析可以帮助学校找到应该集中力量去工作的方向，以确保他们期待的投资得到回报。

与零售行业类似，高等教育机构可以通过制定更具针对性的活动来减少营销支出，从而吸引特定领域的学习者。例如，如果大数据分析显示某些城市的学生经常申请大学，而且经常被录用，那么该机构可以相应地调整其营销工作。

佛罗里达理工大学招生总监 Lauren Willison 在为 IBM 撰写的客座文章中写道："如果大学招生办公室对那些包括最多申请人的入学人员地理位置有着深刻的了解，那么它可以减少营销成本并产生更高的回报率，而不是投资于针对广泛受众群体的营销活动，金钱应该投资于更有可能提高大学收益率的特定市场。"

岗位申请：选择性接收学生

学生提交申请后，大数据分析还不会结束。Fast 公司的投稿人昂格莱德在报道中说，很多机构仍在利用大数据来帮助其决定哪些学生将被录取。对某些数据集的分析可以显示哪些考生最有可能在大学中取得成功，哪些考生有可能辍学或挂科。

威奇塔州立大学目前正在使用这种预测分析方法，帮助管理人员做出更明智的决策。研究表明，这所学校的招生方式可以有 96%的准确率来确定申请人是"高素质的"，或者很可能在该机构做得很好。

识别教育问题

学校也可以使用分析软件来确定哪些学生在教育方面可能存在问题。该策略应用的最好方法之一是在学生学术生涯的早期识别他们可能存在的问题。例如，如果学生在主修课程表现不佳，教师可以在学生落后、挂科或退学之前向他们提供帮助和指导。

威利森指出："学生们可以通过这种解决方案尽早解决课程负担问题，或者在某种程度上探索其他途径，尽早采取措施，而不是辍学。"

从大数据到大财富

随着许多行业都已意识到大数据分析的优势，更多的人正在往这一领域寻求职业发展。 因此，机构正在提供更多的大数据、科学的分析课程和计划。TechRepublic 最近发布了提供这种教育的前 20 名学校的名单，卡内基·梅隆大学、斯坦福大学和圣克拉拉大学名列前茅。

要求和注意事项：需要专家级合作伙伴

大数据分析可以提供无限的机会来改善校园各种流程（如招聘流程、学习流程等）并加强教育。但是，有几点要求管理员应该注意，并不仅仅是数据安全这么简单。

机构必须有一个安全的环境来收集和处理他们的数据，特别是信息中包含学生的敏感个人信息。更重要的是，有许多可能的不同的数据源，必须考虑整合和组织这些数据及其所涉及的工作。

Chapter 10

Big Data in Health

Text A

Big Data in Health Care: Supporting Insights with Colocation

Health care organizations must ensure that any investment in emerging or advanced technology delivers real value for practitioners as well as their patients. Thankfully, this is one of the many sectors in which big data analysis has caught on, helping to bring a range of advantages for doctors, nurses and other providers across the globe.

Healthcare IT News pointed out that big data holds considerable potential for this industry in particular. Through the analysis of depersonalized medical records, data shared from other providers and information from clinical trials, health care practitioners are able to pinpoint individuals who are at risk for certain conditions and ensure they get help. Big data also enables the best possible use of wearable health devices, keeping doctors in the loop about a patient's condition at all times and making sure that treatment intervention takes place at the most opportune time.

As big data analysis continues to grow in the health care sector, so too does colocation. These services enable organizations to utilize space in an expert provider's data center, where servers and other critical computing equipment can be stored, maintained and accessed. In fact, the latest reports show that colocation is

New Words and Expressions

colocation/kɒləʊˈkeɪʃn/ *n.*
主机托管

depersonalize
/diːˈpɜːsənəlaɪz/ *vt.*
使失去个性

intervention/ˌɪntəˈvenʃn/ *n.*
干预；调解

expanding across the board — RnR Market Research reported that the colocation market is on track to reach $54.13 billion by 2020, a significant expansion from its 2015 value of $25.70 billion.

While both big data analysis and colocation can bring numerous benefits for health care institutions, these strategies are even more impactful when leveraged together. Colocation is critical for supporting today's big data initiatives, particularly in the health care industry. Today, we'll take a look at the connection that exists here, and why providers in this marketplace must have colocation support in place to ensure the success of their big data analysis.

Health care and colocation

Colocation has been an attractive option for companies in nearly every marketplace. Leveraging colocation services means that firms need not shell out the capital needed to build, configure, staff and maintain their own dedicated data centers, translating to considerable savings. At the same time, colocation also ensures that organizations' computing equipment — which likely supports some of the most critical databases, applications and platforms required for daily operations — is secured, updated and maintained by a team of experts whose goal is to ensure the top performance of these essential computing assets. With these services in place, a company's own internal IT team isn't bogged down by these responsibilities, and can instead focus on other mission-critical aspects of the business's technology operations.

Colocation is also critical when it comes to compliance. Health care organizations are beholden to several important industry regulations that impact their use of technology. The Health Insurance Portability and Accountability Act (HIPAA) requires that health care providers put certain safeguards in place to ensure the security of electronic protected health information. In other words, all documents containing the sensitive information of patients — medical histories, payment forms, etc. — must be stored and secured in a certain way. Thankfully, there are now expert colocation providers, like Data Realty, that specialize in establishing and maintaining just this type of environment, which helps to guarantee that health care companies are always compliant with the regulations and laws of their industry.

New Words and Expressions

initiative/ɪˈnɪʃətɪv/ *n.* 倡议，主动性

bogged down 停滞不前的

compliance/kəmˈplaɪəns/ *n.* 服从，合规

Bridging the gap: Colocation and big data

Colocation isn't just beneficial for health institutions — it's a must for organizations that want to make the best use of their big data. Colocation support provides everything health care firms need for their analysis initiatives, including:

A centralized repository for data: One of the biggest challenges of any big data project isn't gathering the information, it's ensuring that all of these details are in a single, accessible location and are organized in such a way that they can be utilized by the company's analysis tools. A collocated environment inside an expert provider's data center can offer this centralized location, making it easy for health care organizations to compile, organize and analyze their available information.

Scalable computing resources: Another issue with big data is right in its name sake — the expansive nature of this information requires a certain level of scalability that is difficult to achieve without the help of experts. Thankfully, this is another area in which colocation services shine, offering elastic, scalable resources for health care clients. In this way, the company's data center environment can grow alongside its big data, and customers never have to worry about running out of storage space.

Support for computing-intensive processes: Big data analysis requires the use of specialized tools. However, these programs themselves need a high level of computing support to ensure that analysis processes can take place. Colocation providers can ensure that these demands are met, offering a high-performance computing environment that is ideal for intensive data analysis.

However, when it comes to colocation, not all service providers are the same. Data Realty is a leader in this industry and has partnered with data science expert Aunalytics to create a unified approach to information storage and analysis. Services including data colocation, managed Hadoop hardware clusters and data interconnection ensure that customers can find everything they need in a single location. Best of all, Data Realty and Aunalytics specialize in the needs of the health care industry, offering secure environments for sensitive data.

Note:

The text is adapted from the website:

http://www.linkedin.com/pulse/big-data-health-care-supporting-insights-colocation-short-davis.

New Words and Expressions

institution/ˌɪnstɪˈtʃuːʃən/ *n.*
机构

compile/kəmˈpaɪl/ *v.*
汇编

elastic/ɪˈlæstɪk/ *adj.*
有弹力的

unified/ˈjuːnɪfaɪd/ *adj.*
统一的

Comprehension

Blank Filling

1. Through the analysis of ________ medical records, data shared from other providers and information from clinical trials, health care practitioners are able to pinpoint individuals who are _______ for certain conditions and ensure they get ______.
2. As big data analysis continues to grow in the health care sector, so too does ______.
3. Colocation also ensures that organizations' computing equipment — which likely supports some of the most critical ______, ________ and ________ required for daily operations.
4. A collocated environment inside an expert provider's data center can offer this centralized location, making it easy for health care organizations to ______, _____ and _______ their available information.
5. Services including data ______, managed Hadoop hardware clusters and data ______ ensure that customers can find everything they need in a single location. Best of all, Data Realty and Aunalytics specialize in the needs of the health care industry, offering _________ for sensitive data.

Content Questions

1. What do health care institutions must ensure?
2. What can big data do in the medical field?
3. What is the purpose of HIPAA to require health care providers to develop certain safeguards?
4. What does the colocation service provide for the health care company?

Answers

Blank Filling

1. depersonalized; at risk; help
2. colocation
3. databases; applications; platforms
4. compile; organize; analyze
5. collocation; interconnection; secure environments

Content Questions

1. Health care organizations must ensure that any investment in emerging or advanced technology delivers real value for practitioners as well as their patients.
2. Through the analysis of depersonalized medical records, data shared from other providers and information from clinical trials, health care practitioners are able to pinpoint

individuals who are at risk for certain conditions and ensure they get help.

3. HIPAA requires that health care providers put certain safeguards in place to ensure the security of electronic protected health information. In other words, all documents containing the sensitive information of patients — medical histories, payment forms, etc. — must be stored and secured in a certain way.
4. A centralized repository for data, scalable computing resources, support for computing-intensive processes.

参考译文

医疗保健中的大数据：通过主机托管提供支持

医疗保健机构必须确保在新兴技术或先进技术方面的投资能够为从业者及其患者带来真正的价值。值得庆幸的是，这是大数据分析涉及的众多领域之一，有助于为全球医生、护士和其他供应商提供一系列的帮助。

Healthcare IT News 网站指出，大数据在这个行业的潜力尤为巨大。通过对非个性化的医疗记录、其他提供者共享的数据和临床试验的信息进行分析，医疗从业者能够准确地发现存在健康风险的人，并向他们提供帮助。大数据还可以尽最大可能更好地使用可穿戴的健康设备，使医生随时关注患者的状况，并在最合适的时间进行医疗干预。

随着医疗保健部门的大数据分析以及数据托管的继续发展，这些服务使组织能够利用那些专业供应商的数据中心空间，该空间可以存储、维护和访问服务器和其他关键计算设备。事实上，最新的报告显示，托管业务正在全面扩大——RnR Market Research 报道，到 2020 年，托管市场将达到 541.3 亿美元，相比于 2015 年的 257 亿美元会有大幅增长。

虽然大数据分析和托管都可为医疗保健机构带来许多好处，但是这些策略在一起融合使用时更具影响力。数据托管对于支持今天的大数据计划至关重要，特别是在医疗保健行业。接下来，我们将看看两者的联系，以及为什么这个市场中的服务提供商必须具备托管能力，才能确保大数据分析的成功。

医疗保健和主机托管

主机托管几乎对市场的每个公司都具有吸引力。因为利用托管服务意味着公司不需要支出基础设施构建、配置硬件、管理员和维护自己专用数据中心所需的成本，从而节省了大量资金转化为可见的储蓄。同时，托管可以确保组织的计算设备由专业人士进行更新和维护后保持最佳的性能，这些设备是支持日常运营所需的一些最关键的数据库、应用程序和平台。通过这些服务，公司自己的内部 IT 团队不会因为这些任务而拖延主要任务，而是可以专注于业务技术运营的其他关键方面的任务。

涉及法律法规时，托管也是至关重要的。医疗保健组织受到影响其技术使用的若干重要行业法规的制约。健康保险携带责任法案（HIPAA）要求医疗保健提供者制定一些保障措施，以确保数字健康信息的安全。换句话说，包含患者敏感信息的所有文件——病历、付款表格等必须以某种方式存储和保护。幸运的是，现在有专业托管服务提供商，如Data Realty，专门建立和维护这种类型的环境，这有助于保证医疗保健公司始终遵守行业的规则和法规。

弥补差距：主机托管和大数据

数据托管不仅对医疗机构有益，对于想要充分利用其大数据的组织来说，这也是一项必要措施。数据托管为医疗保健公司的分析行为提供了它们所需的一切帮助，包括：

集中式数据库。任何大型数据项目面临的最大挑战之一不是收集信息，而是确保所有这些数据细节都处于一个单一的可访问的存储空间，并以结构化方式进行组织，以便公司的分析工具利用。专业服务提供商的数据中心可以构造这种共享环境，提供这种集中式的数据存储，使医疗机构易于编辑、组织以及分析其可用信息。

可扩展的计算资源。大数据的另一个问题正如同它的名字——这种信息的广泛性质需要一定程度的可扩展性，在没有专家的帮助下难以实现。幸运的是，这是托管服务的另一个领域，为医疗保健客户提供具有弹性的、可扩展的资源。这样，公司的数据中心环境可以与其大数据一起发展，客户永远不用担心存储空间不足。

支持计算密集型处理。大数据分析需要使用专门的工具。然而，这些程序本身需要高水平的计算支持，以确保分析过程平稳运行。托管提供商可以确保满足这些需求，提供对于密集型数据分析理想的高性能计算环境。

但是，在涉及托管时，并非所有的服务提供商都是一样的。Data Realty是该行业的领导者，并与数据专家Aunalytics（Aunalytics是一个大数据分析公司）合作，共同创建统一的信息存储和分析方法。服务包括数据托管、管理Hadoop的硬件集群和数据互连，确保客户能够在某个位置找到所需的一切。最重要的是，他们能够专注于医疗保健行业的需求，为敏感数据提供安全环境。

Text B

Big Data in healthcare = Big Health?

It was just a little side notice: The pharmaceutical company Sanofi and Verily Life Sciences, belonging to the Big Data giant Google, form a joint venture called Onduo. Together they will support patients with type II in taking the medication timely, adequately and they will raise the patients' awareness for healthy behavior. How exactly they want to achieve that, the published press release does not tell.

Google and Big Data

This joint venture is not the first undertaking of Verily Life Sciences. 2014, the company was working on a contact lens that would measure blood sugar levels via the eye fluid. In 2015 that

New Words and Expressions

pharmaceutical company
制药公司

verily/ˈverəli/ *adv.*
真正地；真实地

joint venture
合资企业

press release
新闻稿

eye fluid
眼液

very company developed a medical identification bracelet to measure heart rate, skin temperature and sunlight. It is clear that it is about data, about Big Data, where Google is involved. And Google is getting ready to become a major player in the health sector too.

The data sea

Big Data in the medical field is a current, but not a new phenomenon. While Germany is in a slumber concerning this issue, the move towards Big Health Data in other countries is well underway. It has been recognized that in the medical field, a large amount of data is obtained: Personal patient data, disease history, family medical history, medical reports, medical expenses, data that may be incurred by technical equipment in the course of a treatment, such as by MRI, blood tests, X-ray, but also self-imposed health data e.g by means of health apps.

Fishing in the sea of data

The analysis of these heterogeneous data sets can reveal previously unknown relationships. What factors favor the emergence of a disease? Who is predominantly affected? What prevention measures are effective? Which therapy is promising for which group of people? At the same time costs can be reduced in the health sector: health insurance can compare courses of treatment for the same symptoms and prevent possible misdiagnosis. Likewise, doctors can decide on therapies based on similar illnesses, hospitals could improve their bed planning. Costly clinical trials could be superfluous.

Dr. Altman and Dr. Tatonetti from Stanford University found out — solely on the basis of data analysis — that the antidepressant Paxil and the cholesterol-lowering drug Pravastatin taken in parallel lead to an increase in blood sugar levels.

For information on risks and side-effects

These are the advantages of Big Health Data. According to data protectionists however there are significant risks too. Medical data is highly sensitive information, which is not intended for foreign eyes. A 100 percent protection of this data cannot be guaranteed since systems can be hacked and sensitive information could get into the wrong hands. Then the question arises, for which aim all the data collected will be used. The fact escapes in many

New Words and Expressions

slumber/ˈslʌmbər/ *n.*
睡觉，睡眠

medical expenses
医疗费用

heterogeneous data
异构数据

predominantly/prɪˈdɒmɪnəntli/ *adv.* 占主导地位地

prevention measures
预防措施

misdiagnosis/ˌmɪsdaɪəgˈnəʊsɪs/ *n.*
误诊

clinical trial
临床试验

superfluous /suːˈpɜːfluəs/ *adj.*
过多的；不必要的

get into the wrong hands
落到不妥当的人手里

cases, contrary to German data protection law, the knowledge of the data donor. Lastly one must always be aware of the fact that Big Data is only about correlations, not about causal relation. Big Data cannot substitute scientific evidence.

A glance in the future

In Germany, merging medical data will remain difficult for the time being. The federal structure of health services and the low cross-linking of health actors such as doctors, health insurances, hospitals, corporate health services etc. make practical implementation hard. However, there are first initiatives to make data accessible for large scale analysis. So it is a question of time before the Big "Health" Data wave spills over to Germany.

Perhaps we could use the remaining time to prepare concrete measures on how to benefit from the advantages of Big "Health" Data without throwing privacy overboard. Here a look at our Swiss neighbors might be interesting. There, the project MIDATA was launched some time ago. It is a nonprofit organization that allows citizens, on a voluntary basis to store their data and to decide for themselves whom they give access to their data and for what purposes. The aim is to end the digital feudalism and to promote a self-determined data handling.

New Words and Expressions

merging/mɜ:dʒɪŋ/ *n.* 合并

for the time being 暂时

feudalism/ ˈfju:dəlɪzəm/ *n.* 封建制度，封建主义

Note:

The text is adapted from the website:

http://www.linkedin.com/pulse/big-data-healthcare-health-ina-brecheis.

参考译文

一则简短的报道称：隶属于大数据巨人谷歌公司的赛诺菲制药公司和 Verily 生命科学公司组建了一家名为 Onduo 的合资企业。它们能提示II型糖尿病患者及时用药，提高患者的健康行为意识。发布的新闻稿并没有说明它们如何达到这一目标。

谷歌和大数据

这个合资企业不是 Verily 生命科学公司的第一项事业。2014 年，该公司正在研究通过测量眼睛的液体来获取血糖水平的隐形眼镜。在 2015 年，该公司开发了一种医疗鉴定手镯来测量心率、皮肤温度和日光强度。很明显，这些都关乎数据，关乎谷歌涉及的大数据。谷歌正准备成为医疗行业的主要参与者。

数据的海洋

医疗领域的大数据看起来相对较新，但实际上已经成为当前的一个趋势，而不是新兴事物。虽然德国在这个问题上处于麻木状态，但其他国家的大医疗数据正在蓬勃发展中。

大家已经认识到，医疗领域正在获取大量数据：患者个人资料、疾病史、家庭病史、医疗报告、医疗费用；技术设备在治疗过程中采集可能发生的数据，例如，通过 MRI、血液检测、X 光检查；还可以通过健康应用程序自行采集健康数据。

在数据海洋中获取有用信息

通过对异构数据集的分析可以揭示以前未知的联系。什么因素导致出现疾病？谁主要受影响？哪种预防措施有效？哪一种治疗对于哪一组人来说是有希望的？同时，医疗部门的成本可以降低：健康保险可以比较相同症状的治疗方案，并防止可能的误诊。同样，医生可以根据类似的疾病来决定治疗，医院可以改善他们的床位规划，同时还可以减少不必要的昂贵的临床试验。

斯坦福大学的 Altman 博士和 Tatonetti 博士发现，仅仅基于数据分析，当抗抑郁药 Paxil 和胆固醇降低药物普伐他汀同时摄取时，就会导致血糖水平升高。

风险和副作用的信息

以上都是大健康数据的优点。但是数据保护主义者认为大数据也存在重大的风险。医疗数据是高度敏感的信息，不能用于其他用途。由于系统可能遭到黑客攻击，敏感信息可能会落入不法分子手中，所以无法保证 100%的数据安全。那么问题就出现了：所有收集的数据将基于什么目的使用？事实上，在许多情况下，数据的来源都是违背德国数据保护法的。最后，必须始终意识到大数据只是相关性关系，而不是因果关系。大数据并不能取代科学证据。

大数据的未来

在德国，医疗数据的整合目前仍将是困难的。医疗服务的结构以及医生、医疗保险、医院、企业医疗服务等医疗从业者交流不畅，使得实际工作变得困难重重。然而，我们在数据分析方面已经做了初步的尝试使数据得以大规模地进行分析。所以大的“健康”数据浪潮涌向德国只是时间问题。

也许我们可以利用剩下的时间来制定具体的措施，研究如何在不冒犯隐私的前提下从海量的健康数据中获益。在这里我们可以借鉴瑞士的经验。在瑞士，MIDATA 项目是在很久以前推出的。这是一个非营利项目，允许公民在自愿的基础上存储他们的数据，并自己决定谁以什么目的可以访问他们的数据。其目的是打破数据的壁垒，推动自主的数据处理。

Chapter 11

Big Data in Banking

Text A

The banking industry has developed in its service delivery and technological innovation. Banking services is a critical component for the day to day activities as most transaction are undertaken through the banking sector. The number of customers served in the banking sector has increased exponentially. Each transaction in the banking sector amount to data creation and collection. The banking industry produces a large volume of data on a day to day activities. The adoption big data analytics of the generated data will revolutionize the banking sector at present and in the future.

Customer segmentation

The banking industry is entitled to a lot of personal information of their customers. The available information has a lot of potentials when utilized by the banking sector effectively. The banks currently can track customer transaction in real time. Through the available information, the bank can segment the customer based on different parameters such as net worth; customer preferred credit card among others. The segmentation of customer enables the bank to customized services and bundle packages that are deemed suitable for the different customer segments with high accuracies. Big data allows summarization of the available information into an actionable data that the bank can leverage.

The segmentation of customers has improved banking industry marketing sector. The bank can now develop a marketing strategy

New Words and Expressions

innovation/ˌɪnəˈveɪʃən/ *n.*
改革，创新

exponentially/ˌekspə'nenʃəlɪ/ *adv.* 以指数方式

revolutionise/ˌrevəˈluːʃənaɪz/ *v.*
使发生革命性剧变

segmentation/ˌsegmenˈteɪʃən/ *n.*
分段

entitle/ɪnˈtaɪtl/ *v.*
给予（某人…）的权利

net worth
净值

that is channeled to particular market niches. The customized marketing strategies have increased market reach in the banking sector and widened the customer base of banks.

Improvement of products and services

The bank can follow the conversation of clients on the digital platforms. The available information is used to determine the different needs of the customers and make them available to them in real time. Through evaluating the services offered by other banks, the company can be able to customize its services so that they are unique and gain competitive advantage. Most of the banks believe that leveraging Big data creates competitive advantage in the banking sector.

Operation efficiency

The banking industry is a fast growing industry with ever increasing expectation of customers. The volume of information gathered in the sector is enormous too and is expected to increase in the future. A significant amount of information is challenging to analyze and simplify in the absence of big data. Implementation of big data analytics ensures that the banking industry databases can store and process the information faster and safer for efficient use. The big data thus enabled improved efficiency through which the data of customers is handled.

The aim of many businesses is to lower the cost of operation and increase the business profitability. The big data adoption in the banking industry ensures the operation cost are reduced. This is through automation of most of the repetitive activities in the bank sector that lower the cost of undertaking such activities. The efficiency of operation is also improved through real-time analysis of information and integration across the bank platform and access to the information from all the bank branches.

Big data in the banking sector provides the bank with real-time information in all the operation levels of the company. There are many indicators put in place to monitor the banking operation. As such, a problem can easily be identifying even before it has a catastrophic effect on the bank operation. Big data analytics in banking helps in reducing technical error that impact on the customers.

New Words and Expressions

niche/nɪtʃ/ *n.*
合适的位置（工作等）；有利可图的缺口，商机

repetitive/rɪˈpetətɪv/ *adj.*
重复的

profitability /ˌprɒfɪtəˈbɪlətɪ/ *n.*
获利（状况）

integration/ˌɪntəˈɡreɪʃ(ə)n/ *n.*
整合，一体化

catastrophic/kəˈtæstrəfik/ *adj.*
灾难的

Big data have been accredited with stimulating innovation. The banking operation succeeds on the basis of innovation which not only improves the efficiency of operation but also gives the banks a competitive advantage. The banking industry has adopted big data to come up with innovation to enhance operation such as the mobile banking.

Risk management

The banking sector is left vulnerable due to the large amount of information that it handles. Fraud is one of the major risks that banks face in its day to day operation. The big data enables monitoring of all the transaction. With increased availability of information, the banks can distinguish a genuine transaction from a fraudulent one, and this has drastically reduced the loses of the bank from fraudulent activities. This is though integrating all bank information in a central place that ensures the security of data.

Cyber security has been one of the major safety issues relating to information handled by banks. The big data provide the organization with real-time information that is able to detect any security breach in its platform. The information available also enable the bank to identify any weak spots in its system, and a meant them before cyber criminals exploit them.

The financial market is now globalized due to technological innovation. A ripple of instability in any one economy can be felt across the globe similar to the 2008/9 financial crisis. The Big data provides the banking industry with the ability to evaluate all factors in the market that may impact their operation and be able to put contingency strategy to protect its operation and the interest of its customers and thus lowering risks.

Future of banking

The adoption of the big data in the banking industry has not yet been fully explored. The expenditure in big data analytics in the future is expected to increase as more and more banks fully adopt big data analytics. There is expected to be more innovation and big data techniques in the banking industry. Banks will have to select the most effective technique that will transform its operation. The phase of the banking industry will change when the industry fully adopts the broad application of the big data.

The customer experience is expected to change in the future. The efficiency of bank operation, real-time sharing of information,

New Words and Expression

accredited/əˈkredɪtɪd/ *adj.* 公认的

vulnerable/ˈvʌlnərəbl/ *adj.* 易受攻击的，易受伤的

fraudulent/ˈfrɔːdʒələnt/ *adj.* 欺骗的，欺诈的

drastically/ˈdrɑːstɪkəlɪ/ *adv.* 大大地，彻底地

Cyber security 网络安全

evaluate/ɪˈvæljueɪt/ *v.* 评估，评价；估值

ripple/ˈrɪpəl/ *n.* 涟漪；微波

contingency /kənˈtɪndʒənsi/ *n.* 可能发生的事

expenditure/ɪkˈspendɪtʃər/ *n.* 全部开支，花费

linking to bank industry to other industry and automation of some function will greatly improve service delivery and customer satisfaction in the banking industry. For sure the future of the banking industry relies on big data analytics.

Note:

The text is adapted from the website:

http://www.linkedin.com/pulse/big-data-inbanking-nikunj-thakkar.

Comprehension

Blank Filling

1. Through the analysis of ________ medical records, data shared from other providers and information from clinical trials, health care practitioners are able to pinpoint individuals who are _______ for certain conditions and ensure they get ______.
2. As big data analysis continues to grow in the health care sector, so too does ______.
3. Colocation also ensures that organizations' computing equipment — which likely supports some of the most critical ______, ________ and _______ required for daily operations.
4. A colocated environment inside an expert provider's data center can offer this centralized location, making it easy for health care organizations to ______, _____ and _______ their available information.
5. Services including data ______, managed Hadoop hardware clusters and data ________ ensure that customers can find everything they need in a single location. Best of all, Data Realty and Aunalytics specialize in the needs of the health care industry, offering _________ for sensitive data.

Content Questions

1. What do health care institutions must ensure?
2. What can big data do in the medical field?
3. What is the purpose of HIPAA to require health care providers to develop certain safeguards?
4. What does the colocation service provide for the health care company?

Answers

Blank Filling

1. depersonalized; at risk; help
2. colocation
3. databases; applications; platforms
4. compile; organize; analyze

5. collocation; interconnection; secure environments

Content Questions

1. Health care organizations must ensure that any investment in emerging or advanced technology delivers real value for practitioners as well as their patients.
2. Through the analysis of depersonalized medical records, data shared from other providers and information from clinical trials, health care practitioners are able to pinpoint individuals who are at risk for certain conditions and ensure they get help.
3. HIPAA requires that health care providers put certain safeguards in place to ensure the security of electronic protected health information. In other words, all documents containing the sensitive information of patients — medical histories, payment forms, etc. — must be stored and secured in a certain way.
4. A centralized repository for data, scalable computing resources, support for computing-intensive processes.

参考译文

银行业在为用户提供服务和技术创新上得到了前所未有的发展。银行服务是人们日常活动的关键组成部分，大多数交易都是通过银行进行的。银行业服务的客户数量急剧增多，其部门的每笔交易也都涉及数据的创建和收集，这些事件每天都产生了大量数据。对产生的数据进行大数据分析将彻底改变现有和未来的银行业。

客户划分

银行业有权获得客户的大量个人信息，然后有效利用这些信息，发挥出其巨大的潜力。目前，银行可以实时监控客户交易。通过现有信息，银行可以根据不同的参数（如净值、客户使用信用卡的偏好等）对客户进行细分，这使银行能够更准确地为不同客户群体定制适合的服务。银行可以利用大数据将可用信息变成可以利用的可操作数据。

将这些客户进行分类，可使银行业营销部门的工作水平得以提高。现在，银行可以制定一个营销策略，并用该策略引导特定的市场。定制的营销策略增加了银行业的市场份额，扩展了银行的客户群。

改进产品和服务

银行可以随时跟客户在数字平台上交流，得到的信息将用于判断客户的不同需求，并将这种需求实时提供给他们。同时，银行可以通过评估其他银行的服务来定制自己独一无二的服务，从而获得竞争优势。大多数银行认为利用大数据可以为其创造优势。

运行效率

银行业是一个快速发展的行业，客户对其期望越来越高。银行收集的信息量也越来越多，而且预计未来会持续增加。在没有大数据分析技术的情况下，对大量的信息进行分析和简化是很困难的。实施大数据分析可确保银行业数据库能够更快、更安全地存储和处理信息，以便更有效地利用信息，由此提高处理客户数据的效率。

许多企业的目标是降低经营成本，提高企业盈利能力。同理，银行业采用了大数据使运营成本降低，通过自动化银行的重复活动来降低活动成本。通过实时分析银行平台的信息，整合所有分行的信息来提高运营效率。

银行的大量客户数据为银行提供了所有业务层面的实时信息，也提供了许多指标来监督银行业。因此，银行业能在巨大危险来临前确定问题所在。大数据分析有助于减少对客户产生影响的技术错误。

大数据在银行业的应用是一项巨大的创新。银行业务的创新，不仅提高了经营效率，而且给银行带来了竞争优势。银行业采取大数据提出了加强手机银行业务等的创新。

风险管理

银行部门由于需要处理大量信息，因此有容易受到攻击的风险。欺诈行为是银行日常运营面临的主要风险之一。大数据可以监控所有的交易。随着信息的可用性的增加，银行可以区分真正的交易和欺诈行为，这大大减少了银行被欺诈而造成的损失。将所有银行信息整合在一个数据中心可以确保数据的安全。

网络安全一直是银行信息处理的主要安全问题。大数据提供能够检测公司平台中安全漏洞的实时信息，还能在网络犯罪分子利用这些信息之前找出系统的漏洞。

由于技术创新，金融市场已经实现了全球化。任何一个经济体的不稳定状态都会波及全球，2008—2009 年的全球金融危机就是一个例子。大数据为银行业提供了评估市场上可能影响其业务的所有因素的能力，并且能够采取应急策略来保护其运营和客户利益，从而降低风险。

银行业的未来

银行业大数据还在发展。随着越来越多的银行采用大数据分析，预计未来大数据分析的支出将会增加。更多的创新和大数据技术将涌进银行业。银行必须选择最有效的技术来改变其运作方式。当大数据在银行业广泛应用时，银行业的发展将会发生翻天覆地的变化。

随着大数据在银行业的发展，客户体验将会发生变化。银行业务效率的提高、实时信息的共享更新、银行业与其他行业的联系、功能的自动化等都将大大提高银行业务服务能力和客户满意度。值得肯定的是，银行业的未来会依赖于大数据分析。

Text B

Big Data: Profitability, Potential and Problems in Banking

More than 70% of banking executives worldwide say customer centricity is important to them. However, achieving greater customer centricity requires a deeper understanding of customer needs. Research from Capgemini indicates that only 37% of customers believe that banks understand their needs and preferences adequately.

The truth is that financial institutions are struggling to profit from ever-increasing volumes of data. Banks are only using a small

New Words and Expressions

centricity/ˈsentrɪsɪtɪ/ *n.* 中心性

portion of this data to generate insights that enhance the customer experience. For instance, research reveals that less than half of banks analyze customers' external data, such as social media activities and online behavior. And only 29% analyze customers' share of wallet, one of the key measures of a bank's relationship with its customers.

Only 37% of banks have hands-on experience with live big data implementations, while the majority of banks are still focusing on pilots and experiments. Capgemini research shows that organizational silos are the single biggest barrier to success with big data. A dearth of analytics talent, high cost of data management, and a lack of strategic focus on big data are also major stumbling blocks.

Customer data typically resides in silos across lines of business or is distributed across systems focused on specific functions such as CRM, portfolio management and loan servicing. As such, banks lack a seamless 360-degree view of the customer. Further, many banks have inflexible legacy systems that impede data integration and prevent them from generating a single view of the customer.

Lack of Strategic Focus: Big Data Viewed as Just Another "IT Project"

Big data requires new technologies and processes to store, organize, and retrieve large volumes of structured and unstructured data. Traditional data management approaches followed by banks do not meet big data requirements. For instance, traditional approaches hinge on a relational data model where relationships are created inside the system and then analyzed. However, with big data, it is difficult to establish formal relationships with the variety of unstructured data that comes through. Similarly, most traditional data management projects view data from a static and/or historic perspective.

While most IT projects are driven by the twin facets of stability and scale, big data demands discovery, ability to mine existing and new data, and agility. Consequently, by taking a traditional IT-based approach, organizations limit the potential of big data. In fact, Capgemini says an average company sees a return of just 55 cents on every dollar that it spends on big data.

New Words and Expressions

portion/ˈpɔːʃən/ *n.*
一部分；一份

hands-on
实际动手操作的

dearth/dɜːθ/ *n.*
缺乏

stumbling blocks
绊脚石

seamless/ˈsiːmləs/ *adj.*
无缝的

reside in
存在于，属于

portfolio management
证券管理

loan servicing
贷款服务

inflexible/ɪnˈfleksəbəl/ *adj.*
不可改变的，不愿变更的

legacy/ˈlegəsi/ *n.*
历史遗产

retrieve/rɪˈtriːv/ *v.*
检索

hinge on
取决于…

Deutsche Bank's Big Data Plans Held Back By Legacy Infrastructure

Deutsche Bank has been working on a big data implementation since the beginning of 2012 in an attempt to analyze all of its unstructured data. However, problems have arisen while attempting to unravel the traditional systems — mainframes and databases, and trying to make big data tools work with these systems.

The bank has been collecting data from the front end (trading data), the middle (operations data) and the back-end (finance data). Petabytes of this data are stored across 46 data warehouses, where there is 90% overlap of data. It is difficult to unravel these data warehouses that have been built over the last two to three decades. The data integration challenge and the significant investments made by the bank in traditional IT infrastructure pose a key question for the bank's senior executives — what do they do now with their traditional system? They believe that big, unstructured and raw data analysis will provide important insights, mainly unknown to the bank. But they need to extract this data, streamline it and build traceability and linkages from the traditional systems, which is an expensive proposition.

Reality Check: If a bank the size of Deutsche — one of the biggest banks on earth — struggles with big data, you can be sure that most smaller institutions will face even greater obstacles.

How Can Banks Realize Greater Value From Their Data?

Customer data analytics has been a relatively low priority area for banks. Most have concentrated their energy on risk management, not using analytics to enhance the customer experience, Capgemini says.

But their research shows that banks applying analytics to customer data have a four-percentage point lead in market share over banks that do not. The difference in banks that use analytics to understand customer attrition is even more stark at 12-percentage points.

Capgemini believes banks can maximize the value of their customer data by leveraging big data analytics across the three key areas of customer retention, market share growth and increasing share of wallet.

New Words and Expressions

Deutsche Bank
德意志银行

mainframe/ˈmeɪnfreɪm/ *n.*
（大型计算机的）主机，大型机

overlap/ˌəʊvəˈlæp/ *v.*
（与…）交叠

unravel/ʌnˈrævəl/ *v.*
拆散

stark/stɑːk/ *adv.*
明显地

Maximizing Lead Generation

Big data solutions can help banks generate leads for customer acquisition more effectively. Take the case of US Bank, the fifth largest commercial bank in the US. The bank wanted to focus on multi-channel data to drive strategic decision-making and maximize lead conversions. The bank deployed an analytics solution that integrates data from online and offline channels and provides a unified view of the customer. This integrated data feeds into the bank's CRM solution, supplying the call center with more relevant leads. It also provides recommendations to the bank's web team on improving customer engagement on the bank's website. As a result, the bank's lead conversion rate has improved by over 100% and customers receive an enhanced and personalized experience. The bank also executed three major website redesigns in 18 months, using data-driven insights to refine website content and increase customer engagement.

Next Best Action Analytics Models Unlock Opportunities to Drive Top Line Growth

From "next best offer" to cross-selling and up-selling, the insights gleaned from big data analytics allows financial marketers to make more accurate decisions. Big data analytics allows banks to target specific micro customer segments by combining various data points such as past buying behavior, demographics, sentiment analysis from social media along with CRM data. This helps improve customer engagement, experience and loyalty, ultimately leading to increased sales and profitability.

Predictive analytics can improve conversion rates by seven times and top-line growth ten-fold. Capgemini studied the impact of using advanced, predictive analytics on marketing effectiveness for a leading European bank. The bank shifted from a model where it relied solely on internal customer data in building marketing campaigns to one where it merged internal and external data sets and applied advanced analytics techniques to this combined data set. As a result of this shift, the bank was able to identify and qualify its target customers better.

Big data Analytics Helps Banks Limit Customer Attrition

A mid-sized European bank used data sets of over 2 million customers with over 200 variables to create a model that predicts

New Words and Expressions

shift/ʃɪft/ *v.*
改变

solely/ˈsəʊlli/ *adv.*
单独地

merge/mɜːdʒ/ *v.*
融入

the probability of churn for each customer. An automated scorecard with multiple logistic regression models and decision trees calculated the probability of churn for each customer. Through early identification of churn risks, the bank saved itself millions of dollars in outflows it otherwise could not have avoided.

New Words and Expressions

logistic regression model
逻辑回归模型

decision tree
决策树

outflow/ˈaʊtfləʊ/ *n.*
外流；流出

Note:

The text is adapted from the website:

https://thefinancialbrand.com/38801/big-data-profitability-strategy-analytics-banking/.

参考译文

银行业大数据的盈利能力、潜力和存在问题

全球超过70%的银行业高管认为，深入地了解客户的需求、以客户为中心是至关重要的。但Capgemini的研究表明，只有37%的客户认为银行充分了解他们的需求和偏好。

事实上，金融机构正在从不断增加的数据中获利。银行目前只使用这些数据的一小部分来增强客户的体验。研究显示，不到一半的银行分析客户的外部数据，如社交媒体活动和在线行为。客户资金的分配是衡量银行与客户关系的关键措施，但是只有29%的银行对客户资金进行了分析。

只有37%的银行有实时大数据的实践经验，而大多数银行仍然处于摸索阶段。Capgemini研究表明，分析人才不足、数据管理成本高、缺乏对大数据的战略重点认识度以及组织之间的信息不互通，都成为成功的最大障碍。

客户数据通常存储于业务线上的数据孤岛中，或分布在专注于特定功能的系统中（如CRM、投资组合管理和贷款服务）。因此，银行对客户无法全面了解。此外，许多银行都有不灵活的遗留系统，这阻碍了数据集成并阻止他们生成单一的客户视图。

缺乏战略重点：大数据被视为另一个“IT项目”

大数据需要新的技术和流程，对大量的结构化和非结构化数据进行存储、组织和检索。银行遵循的传统数据管理方法与大数据的管理方法不符。例如，传统方法取决于关系数据模型，数据关系在系统内部建立，然后才进行分析。然而，使用大数据，很难通过各种非结构化的数据建立正式的关系。另外，大多数传统的数据管理项目是从静态或历史的角度来查看数据的，这都与大数据的管理方法不符。

大多数信息技术项目都是由稳定性和规模两方面驱动的，但大数据要求探索和发现，即挖掘现有的数据和发现新数据的能力以及敏捷性。因此，采用传统的基于IT的方法，限制了大数据的潜力。Capgemini表示，一家普通公司的数据表明：在大数据上每一美元的支出，对应只有55美分的收益。

德意志银行的大数据计划以传统基础设施为背景

自 2012 年年初以来，德意志银行一直在开展大数据分析的实施方案，试图分析出所有非结构化数据。然而，在尝试研究传统系统大型机和数据库，并使大数据工具与这些系统一起工作时出现了问题。

传统银行从前端（交易数据）、中端（操作数据）和后端（财务数据）收集数据。这些 PB 级数据存储在 46 个 90%重叠的数据库中。这些数据库已经建立了二三十年，所以很难被分解开。银行高管人员认为，数据集成的挑战以及银行在传统 IT 基础设施方面的重大投资是目前的关键问题——他们现在用传统的系统做什么？他们认为非结构化的原始数据给银行提供了重要的信息。他们需要提取并简化这些数据，构建可追溯的与传统系统的联系，显然目前这个计划是非常昂贵的。

现实：德意志银行（世界上最大的银行之一）仍然视大数据为待攻克的难题，所以大多数较小的银行同样面临更大的障碍。

银行如何从数据中实现更大的价值？

对于银行业，客户数据分析和其他业务相比并未受到同等的重视。Capgemini 指出，大多数银行将精力集中在风险管理上，而不是使用分析来增强客户体验。

研究显示，对客户数据进行分析的银行所占的市场份额比没有分析的银行要高出四个百分点。与使用了数据来分析客户消费的银行之间的差异就更加明显，为 12 个百分点。

Capgemini 认为，银行可以通过在三个关键领域利用大数据分析来最大限度地发挥客户数据的价值：客户的留存率、市场份额的增长和金钱份额的增加。

大限度地提高数据的主导性

大数据解决方案可帮助银行更有效地获取潜在客户。以美国第五大商业银行为例，该银行希望通过专注于多渠道数据，推动战略决策，最大限度地提高访客成交率。银行部署了一个分析解决方案，用以整合来自在线和离线渠道的数据，并提供客户的统一视图。该数据可以给银行的 CRM（客户关系管理）提供解决问题的方案，为呼叫中心提供更多相关的潜在客户。它还向银行网络团队提供了关于改善银行网站上客户参与度的建议。因此，银行的访客成交率提高了 100%以上，客户也获得增强的、个性化的体验。该银行还在 18 个月内重新设计了三个主要网站，使用数据驱动等方法来改进网站内容，增加了客户参与度。

采用最佳行为分析模型将开启推动收入增长的机会

从“最佳报价”到交叉销售和向上销售，金融营销人员通过大数据分析做出了准确的决策。通过结合各种数据点（例如过去的购买行为、人口统计、社交媒体的情绪分析以及 CRM 数据）来针对特定的微型客户群体。这有助于提高客户参与度、体验和忠诚度，最终使销售水平和盈利能力得到了提高。

预测分析可以将成交率提高了七倍，收入增长了十倍。Capgemini 研究了使用预测分析对欧洲领先银行的营销效果的影响。银行从一个完全依靠内部客户数据建立营销活动的模式转换到一个将内部和外部数据集合并应用于高级分析技术的组合数据集中模式中。由于这一转变，银行能够更好地识别和评估其目标客户。

大数据分析可减少银行的客户流失

一家中等规模的欧洲银行使用了包括超过 200 万个客户的数据集，有 200 多个变量来创建一个预测客户流失概率的模型，其中一个自动记分卡可计算每个客户流失的概率，它是由多个逻辑回归模型和决策树搭建。银行也通过大数据分析及时识别出客户流失风险因素，从而避免数百万美元的资金外流。

附录 A

常用大数据词汇中英文对照表

A

Aggregation	聚合——搜索、合并、显示数据的过程
Algorithms	算法——可以完成某种数据分析的数学公式
Analytics	分析法——用于发现数据的内在含义
Anomaly detection	异常检测——在数据集中搜索与预期模或行为不匹配的数据项。除 anomalies，用来表示异常的词有以下几种：outliers，exceptions，surprises，contaminants。它们通常可提供关键的可执行信息
Anonymization	匿名化——使数据匿名，即移除所有与个人隐私相关的数据
Application	应用——实现某种特定功能的计算机软件
Artificial intelligence	人工智能——研发智能机器和智能软件，这些智能设备能够感知周遭的环境，并根据要求作出相应的反应，甚至能自我学习

B

Behavioural analytics	行为分析法——这种分析法是根据用户的行为如“怎么做”“为什么这么做”以及“做了什么”来得出结论，而不是仅仅针对人物和时间的一门分析学科，它着眼于数据中的人性化模式
Big data scientist	大数据科学家——能够设计大数据算法、使得大数据变得有用的人
Big data startup	大数据创业公司——指研发最新大数据技术的新兴公司
Biometrics	生物测定术——根据个人的特征进行身份识别
BB, Brontobytes	B 字节——约等于 1000YB（Yottabytes），相当于未来数字化宇宙的大小
Business intelligence	商业智能——一系列理论、方法学和过程，使得数据更容易被理解

C

Classification analysis	分类分析——从数据中获得重要的相关性信息的系统化过程；

	这类数据也被称为元数据（meta data），是描述数据的数据
Cloud computing	云计算——构建在网络上的分布式计算系统，数据是存储于机房外的（即云端）
Clustering analysis	聚类分析——将相似的对象聚合在一起，每类相似的对象组合成一个聚类（也叫作簇）的过程。这种分析方法的目的在于分析数据间的差异和相似性
Cold data storage	冷数据存储——在低功耗服务器上存储那些几乎不被使用的旧数据。这些数据检索起来将会很耗时
Comparative analysis	对比分析——在非常大的数据集中进行模式匹配时，进行一步步地对比和计算过程，得到分析结果
Complex structured data	复杂结构的数据——由两个或多个复杂而相互关联部分组成的数据，这类数据不能简单地由结构化查询语言或工具（SQL）解析
Computer generated data	计算机产生的数据——如日志文件这类由计算机生成的数据
Concurrency	并发——同时执行多个任务或运行多个进程
Correlation analysis	相关性分析——一种数据分析方法，用于分析变量之间是否存在正相关或者负相关
CRM	客户关系管理（customer relationship management）——用于管理销售、业务过程的一种技术，大数据将影响公司的客户关系管理的策略

D

Dashboard	仪表板——使用算法分析数据，并将结果用图表方式显示于仪表板中
Data aggregation tools	数据聚合工具——将分散于众多数据源的数据转化成一个全新数据源的过程
Data analyst	数据分析师——从事数据分析、建模、清理、处理的专业人员
Database	数据库——一个以某种特定的技术来存储数据集合的仓库
Database-as-a-service	数据库即服务——部署在云端的数据库，即用即付，例如亚马逊云服务（Amazon Web Services，AWS）
DBMS	数据库管理系统（database management system）——收集、存储数据，并提供数据的访问
Data centre	数据中心——一个实体地点，放置用来存储数据的服务器
Data cleansing	数据清洗——对数据进行重新审查和校验的过程，目的在于删除重复信息、纠正存在的错误，并提供数据一致性
Data custodian	数据管理员——负责维护数据存储所需技术环境的专业技术人员
Data ethical guidelines	数据道德准则——这些准则有助于组织机构使其数据透明化，保证数据的简洁、安全及隐私
Data feed	数据订阅——一种数据流，例如 Twitter 订阅和 RSS

Data marketplace　数据集市——进行数据集买卖的在线交易场所

Data mining　数据挖掘——从数据集中发掘特定模式或信息的过程

Data modeling　数据建模——使用此技术来分析数据对象，以此洞悉数据的内在含义

Data set　数据集——大量数据的集合

Data virtualization　数据虚拟化——数据整合的过程，以此获得更多的数据信息，这个过程通常会引入其他技术，例如数据库、应用程序、文件系统、网页技术、大数据技术等

De-identification　去身份识别——也称为匿名化（anonymization），确保个人不会通过数据被识别

Discriminant analysis　判别分析——将数据分类；按不同的分类方式，可将数据分配到不同的群组、类别或者目录。是一种统计分析法，可以对数据中某些群组或集群的已知信息进行分析，并从中获取分类规则

Distributed File System　分布式文件系统——提供简化的、高可用的方式来存储、分析、处理数据的系统

Document store databases　文件存储数据库——又称为文档数据库（document-oriented database），为存储、管理、恢复文档数据而专门设计的数据库，这类文档数据也称为半结构化数据

E

Exploratory analysis　探索性分析——在没有标准流程或方法的情况下从数据中发掘模式，是发掘数据和数据集主要特性的一种方法

EB, Exabytes　E 字节——约等于 1000PB（Petabytes），约等于 10^9GB。如今全球每天所制造的新信息量大约为 1EB

ETL，extract, transform and load　提取-转换-加载——一种用于数据库或者数据仓库的处理过程。即从各种不同的数据源提取（E）数据，并转换（T）成能满足业务需要的数据，最后将其加载（L）到数据库

F

Failover　故障切换——当系统中某个服务器发生故障时，能自动地将运行任务切换到另一个可用服务器或节点上

Fault-tolerant design　容错设计——一个支持容错设计的系统应该能够做到当某一部分出现故障也能继续运行

G

Gamification　游戏化——在非游戏领域中运用游戏的思维和机制，这种方法可以一种十分友好的方式进行数据的创建和侦测，非常有效

Graph databases　图形数据库——运用图形结构（例如一组有限的有序对，或者某种实体）来存储数据。这种图形存储结构包括边缘、属性和节点，它提供了相邻节点间的自由索引功能，也就是说，数据库中每个元素都与其他相邻元素直接关联

Grid computing	网格计算——将许多分布在不同地点的计算机连接在一起，用以处理某个特定问题，通常是通过云将计算机相连在一起
H	
Hadoop	一个开源的分布式系统基础框架，可用于开发分布式程序，进行大数据的运算与存储
HBase	Hadoop 数据库——一个开源的、非关系型、分布式数据库，与 Hadoop 框架共同使用
HDFS	Hadoop 分布式文件系统（Hadoop Distributed File System）——一个被设计成适合运行在通用硬件（commodity hardware）上的分布式文件系统
HPC	高性能计算（high-performance-computing）——使用超级计算机来解决极其复杂的计算问题
I	
IMDB	内存（in-memory）数据库——一种数据库管理系统，与普通数据库管理系统不同之处在于，它用主存来存储数据，而非硬盘。其特点在于能高速地进行数据的处理和存取
Internet of Things	物联网——在普通的设备中装上传感器，使这些设备能够在任何时间任何地点与网络相连
J	
Juridical data compliance	法律上的数据一致性——当使用的云计算解决方案将数据存储于不同的国家或不同的大陆时，就会与这个概念扯上关系。需要留意这些存储在不同国家的数据是否符合当地的法律
K	
Key value databases	键值数据库——数据的存储方式是使用一个特定的键，指向一个特定的数据记录，这种方式使得数据的查找更加方便快捷。键值数据库中所存的数据通常为编程语言中基本数据类型的数据
L	
Latency	延迟——表示系统时间的延迟
Legacy system	遗留系统——一种旧的应用程序，或是旧的技术，或是旧的计算系统，现在已经不再支持
Load balancing	负载均衡——将工作量分配到多台计算机或服务器上，以获得最优结果和最大的系统利用率
Location data	位置信息—— GPS信息，即地理位置信息
Log file	日志文件——由计算机系统自动生成的文件，记录系统的运行过程
M	
Machine 2 machine data	M2M 数据——两台或多台机器间交流与传输的内容

Machine data	机器数据——由传感器或算法在机器上产生的数据
Machine learning	机器学习——人工智能的一部分，指的是机器能够从它们所完成的任务中进行自我学习，通过长期的累积实现自我改进
MapReduce	处理大规模数据的一种软件框架（Map: 映射，Reduce: 归纳）
MPP	大规模并行处理（massively parallel processing）——同时使用多个处理器（或多台计算机）处理同一个计算任务
Metadata	元数据——被称为描述数据的数据，即描述数据属性（数据是什么）的信息
MongoDB	一种开源的非关系型数据库（NoSQL database）
Multi-dimensional databases	多维数据库——用于优化数据联机分析处理（OLAP）程序、优化数据仓库的一种数据库
MultiValue databases	多值数据库——一种非关系型数据库（NoSQL），一种特殊的多维数据库，能处理 3 个维度的数据。主要针对非常长的字符串，能够完美地处理 HTML 和 XML 中的字符串

N

Natural language processing	自然语言处理——计算机科学的一个分支领域，它研究如何实现计算机与人类语言之间的交互
Network analysis	网络分析——分析网络或图论中节点间的关系，即分析网络中节点间的连接和强度关系
NewSQL	一个优雅的、定义良好的新型数据库系统，比 SQL 更易学习和使用，比 NoSQL 更晚提出
NoSQL	顾名思义，就是“不使用 SQL”的数据库，泛指传统关系型数据库以外的其他类型的数据库。这类数据库有更强的一致性，能处理超大规模和高并发的数据

O

Object databases	对象数据库——也称为面向对象数据库，以对象的形式存储数据，用于面向对象编程。它不同于关系型数据库和图形数据库，大部分对象数据库都提供一种查询语言，允许使用声明式编程（declarative programming）访问对象
Object-based image analysis	基于对象的图像分析——数字图像分析方法是对每一个像素的数据进行分析，而基于对象的图像分析方法则只分析相关像素的数据，这些相关像素被称为对象或图像对象
Operational databases	操作型数据库——这类数据库可以完成一个组织机构的常规操作，对商业运营非常重要，一般使用在线事务处理，允许用户访问、收集、检索公司内部的具体信息
Optimization analysis	优化分析——在产品设计周期依靠算法来实现的优化过程，在这一过程中，公司可以设计各种各样的产品并测试这些产品是否满足预设值

Ontology　本体论——表示知识本体，用于定义一个领域中的概念集及概念之间的关系的一种哲学思想。数据被提高到哲学的高度，被赋予了世界本体的意义，成为一个独立的客观数据世界

Outlier detection　异常值检测——异常值是指严重偏离一个数据集或一个数据组合总平均值的对象，该对象与数据集中的其他相去甚远，因此，异常值的出现意味着系统发生问题，需要对此另加分析

P

Pattern recognition　模式识别——通过算法来识别数据中的模式，并对同一数据源中的新数据作出预测

PB, Petabyte　P 字节——约等于 1000TB（Terabytes），约等于 100 万 GB（Gigabytes）。欧洲核子研究中心（CERN）大型强子对撞机每秒产生的粒子个数就约为 1PB

PaaS　平台即服务（Platform-as-a-Service）——为云计算解决方案提供所有必需的基础平台的一种服务

Predictive analysis　预测分析——大数据分析方法中最有价值的一种分析方法，这种方法有助于预测个人未来（近期）的行为（例如，某人很可能会买某些商品，可能会访问某些网站、做某些事情或者产生某种行为）。通过使用各种不同的数据集，如历史数据、事务数据、社交数据或者客户的个人信息数据，来识别风险和机遇

Privacy　隐私——把具有可识别出个人信息的数据与其他数据分离开，以确保用户隐私

Public data　公共数据——由公共基金创建的公共信息或公共数据集

Q

Quantified self　数字化自我——使用应用程序跟踪用户一天的一举一动，从而更好地理解其相关的行为

Query　查询——查找某个问题答案的相关信息

R

Re-identification　再识别——将多个数据集合并在一起，从匿名化的数据中识别出个人信息

Regression analysis　回归分析——确定两个变量间的依赖关系。这种方法假设两个变量之间存在单向的因果关系（自变量和因变量，二者不可互换）

RFID　射频识别——这种识别技术使用一种无线非接触式射频电磁场传感器来传输数据

Real-time data　实时数据——指在几毫秒内被创建、处理、存储、分析并显示的数据

Recommendation engine　推荐引擎——推荐引擎算法根据用户之前的购买行为或其他购买行为向用户推荐某种产品

Routing analysis	路径分析——针对某种运输方法，通过使用多种不同的变量分析从而找到一条最优路径，以达到降低燃料费用、提高效率的目的
S	
Semi-structured data	半结构化数据——半结构化数据并不具有结构化数据严格的存储结构，但它可以使用标签或其他形式的标记方式以保证数据的层次结构
Sentiment analysis	情感分析——通过算法分析出人们是如何看待某些话题
Signal analysis	信号分析——通过度量随时间或空间变化的物理量来分析产品的性能，特别是使用传感器数据
Similarity searches	相似性搜索——在数据库中查询最相似的对象，这里所说的数据对象可以是任意类型的数据
Simulation analysis	仿真分析——仿真是指模拟真实环境中进程或系统的操作，仿真分析可以在仿真时考虑多种不同的变量，确保产品性能达到最优
Smart grid	智能网格——在能源网中使用传感器实时监控其运行状态，有助于提高效率
SaaS	软件即服务（Software-as-a-Service）——基于 Web 的、通过浏览器使用的一种应用软件
Spatial analysis	空间分析——空间分析法分析地理信息或拓扑信息这类空间数据，从中得出分布在地理空间中的数据的模式和规律
SQL	在关系型数据库中，用于检索数据的一种编程语言
Structured data	结构化数据——可以组织成行列结构，可识别的数据。这类数据通常是一条记录，或者一个文件，或者被正确标记过的数据中的某一个字段，并且可以被精确地定位到
T	
TB，Terabyte	T 字节 ——约等于 1000GB（Gigabytes）。1TB 容量可以存储约 300 小时的高清视频
Time series analysis	时序分析——分析在重复测量时间里获得的定义良好的数据。分析的数据必须是良好定义的，并且要取自相同时间间隔的连续时间点
Topological data analysis	拓扑数据分析——拓扑数据分析主要关注三点：复合数据模型、集群的识别以及数据的统计学意义
Transactional data	交易数据——随时间变化的动态数据
Transparency	透明性——消费者想要知道他们的数据有什么作用、被如何处理，而组织机构则把这些信息都透明化了
U	
Un-structured data	非结构化数据——非结构化数据一般被认为是大量纯文本数据，其中还可能包含日期、数字和实例

V

Value	价值——大数据 4V 特点之一。所有可用的数据能为组织机构、社会、消费者创造出巨大的价值。这意味着各大企业及整个产业都将从大数据中获益
Variability	可变性——也就是说，数据的含义总是在（快速）变化的。例如，一个词在相同的推文中可以有完全不同的意思
Variety	多样——大数据 4V 特点之一。数据总是以各种不同的形式呈现，如结构化数据、半结构化数据、非结构化数据，甚至还有复杂结构化数据
Velocity	高速——大数据 4V 特点之一。在大数据时代，数据的创建、存储、分析、虚拟化都要求被高速处理
Veracity	真实性——组织机构需要确保数据的真实性，才能保证数据分析的正确性。因此，真实性是指数据的正确性
Visualization	可视化——只有正确地可视化，原始数据才可被投入使用。这里的“可视化”指的并非普通的图形或饼图，而是复杂的图表，图表中包含大量的数据信息，但可以被很容易地理解和阅读
Volume	大量——大数据 4V 特点之一。指数据量，范围从 Megabytes 至 Brontobytes

W

Weather data	天气数据——是一种重要的开放公共数据来源，如果与其他数据来源合成在一起，可以为相关组织机构提供深入分析的依据

X

XML databases	XML 数据库——XML 数据库是一种以 XML 格式存储数据的数据库。XML 数据库通常与面向文档型数据库相关联，开发人员可以对 XML 数据库的数据进行查询，导出以及按指定的格式序列化

Y

YB,Yottabyte	Y 字节——约等于 1000ZB（Zettabytes），约等于 250 万亿张 DVD 的数据容量

Z

ZB,Zettabyte	Z 字节——约等于 1000EB（Exabytes），约等于 10^9 TB

附录 B

存储容量单位换算

8 bits = 1 Byte（字节）
1024 Bytes = 1 Kilobyte
1024 Kilobytes = 1 Megabyte
1024 Megabytes = 1 Gigabyte
1024 Gigabytes = 1 Terabyte
1024 Terabytes = 1 Petabyte
1024 Petabytes = 1 Exabyte
1024 Exabytes = 1 Zettabyte
1024 Zettabytes = 1 Yottabyte
1024 Yottabytes = 1 Brontobyte
1024 Brontobytes = 1 Geopbyte

参 考 文 献

[1] 孟小峰，慈祥. 大数据管理：概念、技术与挑战[J]. 计算机研究与发展，2013,50(1): 146-169.

[2] 刘雅辉，张铁赢，靳小龙，等. 大数据时代的个人隐私保护[J]. 计算机研究与发展，2015, 52(1): 229-247.

[3] 袁露，肖志勇，王映龙. 论大数据的现状及其发展研究[J]. 教育教学论坛，2014(44): 86-87.

[4] 秦萧，甄峰. 基于大数据应用的城市空间研究进展与展望[D]. 南京：南京大学，2013.

[5] 荆林波. 大数据时代带来的大变革[N/OL]. 中国青年报，2014-05-26(2). http://zqb.cyol.com/html/2014-05/26/nw.D110000zgqnb_20140526_2-02.htm.

[6] 张颖超. 大数据对高等教育发展的影响研究[D]. 重庆：重庆大学，2016.

[7] 伊恩・艾瑞斯. 大数据思维与决策[M]. 北京：人民邮电出版社，2014.

[8] Perera C, Ranjan R, Wang L, et al. Big data privacy in the Internet of Things Era[J]. IT Professional, 2015, 17(3): 32-39.

[9] Talia D. Cloud for Scalable Big Data Analytics[J]. Computer, 2013, 46(5): 98-101.

[10] Erdman A G, Keefe D F, Schiestl R. Grand Challenge: Applying Regulatory Science and Big Data to Improve Medical Device Innovation[J]. IEEE transactions on bio-medical engineering, 2013, 60(3): 589-602.

[11] Dobre C, Xhafa F. Intelligent services for Big data science[J]. Future generations computer systems: FGCS, 2014, 37(2): 267-281.

[12] Ma Y, Wu H, Wang L, et al. Remote sensing big data computing: Challenges and opportunities[J]. Future generations computer systems: FGCS, 2015, 51: 47-60.

[13] Spiess J, Joens Y T, Dragnea R, et al. Using Big data to Improve Customer Experience and Business Performance[J]. Bell Labs Technical Journal, 2014, 18(4): 3-17.

[14] Zwick M. Big data in Official Statistics[J]. Bundesgesundheitsblatt Gesundheitsforschung Gesundheitsschutz, 2015, 58(8): 838-843.

[15] Mayer-Schönberger V. Big data: A revolution that will transform our lives[J]. Bundesgesundheitsblatt Gesundheitsforschung Gesundheitsschutz, 2015, 58(8): 788-793.

[16] Kari V, Geetha M A. Review on Big Data & Analytics — Concepts, Philosophy, Process and Applications[J]. Cybernetics and Information Technologies, 2017, 17(2): 3-27.

[17] Victor C, Muthu R, Gary W, et al. Editorial for FGCS special issue: Big Data in the cloud[J]. Future Generation Computer Systems, 2016, 65: 73-75.

[18] Mazzei M J, Noble D. Big data dreams: A framework for corporate strategy[J]. Business Horizons, 2017, 60(3): 405-414.

[19] Tesfagiorgish D G, Li J Y. Big Data Transformation Testing Based on Data Reverse Engineering[J]. Ubiquitous Intelligence & Computing & IEEE, 2016: 649-652.

[20] Li M, Zhu L. Big Data Processing Technology Research and Application Prospects[C]. Harbin: International Conference on Instrumentation & Measurement, 2014, 7(2): 269-273.

[21] Bhardwaj V, Johari R. Big Data Analysis: Issues and Challenges[C]. Visakhapatnam: International Conference on Electrical, 2015: 1-6.

[22] Benjelloun F Z, Lahcen A A, Belfkih S. An overview of big data opportunities, applications and tools[C]. Fez: Intelligent Systems & Computer Vision, 2015: 1-6.

[23] E XH, Han J, Wang Y S, et al. Big Data-as-a-Service:Definition and architecture[C]. Guilin: IEEE International Conference on Communication Technology, 2013: 738-742.

[24] James M, Michael C, Brad B, et al. Big data: The next frontier for innovation, competition and productivity. Mckinsey Global Institute [N/OL]. 2011. http://www.mckinsey.com/business-functions/digital-mckinsey/our-insights/ big-data-the-next-frontier-for-innovation.

[25] SINTEF. Big Data, for better or worse: 90% of world's data generated over last two years. Science Daily [N/OL]. 2013. https://www.sciencedaily.com/releases/2013/05/130522085217.htm.

[26] Mayer-Schönberger V, Cukier K. Big Data: A Revolution That Will Transform How We Live, Work, and Think[M]. Boston: Eamon Dolan/Houghton Mifflin Harcourt, 2013.

[27] Patil M R, Thia F. Pentaho for Big Data Analytics[M]. Birmingham: Packt Publishing, 2013.